Basic
Arc Welding
(SMAW)

Basic
Arc Welding
(SMAW)

FOURTH EDITION

IVAN H. GRIFFIN • EDWARD M. RODEN •
CHARLES W. BRIGGS

DELMAR PUBLISHERS INC.®

Cover photo by Larry Jeffus

Administrative editor: Mark Huth
Production editor: Peter Hornik

For information address Delmar Publishers Inc.
2 Computer Drive West, Box 15-015
Albany, New York 12212

Printed in the United States of America
Published simultaneously in Canada
by Nelson Canada,
A division of The Thomson Corporation

10 9 8 7

Library of Congress Cataloging in Publication Data

Griffin, Ivan H.
    Basic arc welding.

    Includes index.
    1. Electric welding. I. Roden, Edward M. II. Briggs,
Charles W. III. Title.
TK4660.G743   1984   671.5'212   83-18843
ISBN 0-8273-2131-7

CONTENTS

CHARTS

PREFACE

Shielded metal arc welding (SMAW) is a popular and efficient method of joining metals. This welding process has undergone much development in recent years, especially in the areas of electrodes, machines, and related equipment. BASIC ARC WELDING (SMAW) helps the beginning welder develop skill in this process.

This basic textbook acquaints the beginning welder with this method. Most of the units include step-by-step procedures for performing basic welds on steel plates. Each unit explores a new aspect of welding with an electric arc. The first few units are introductory, presenting the necessary background knowledge. After this introductory section, each "hands-on" unit requires a greater degree of eye-hand coordination which will be developed through experimentation. All of the units include review questions to check the student's progress. So that the student can see the effect of varying the procedure or manipulation of the equipment, experimentation is encouraged.

This revision of BASIC ARC WELDING makes particular use of pictures of welds being made so that students can visualize what is required. No previous skill, knowledge, or training in welding is required for the student or instructor to use this textbook. All of the unit material, both new and old, has been reviewed for readability and reliability. The addition of new illustrations helps update the book. Also, an index has been added to give the student fast and easy reference.

Because of its clear and readable format, this text has proved popular in prevocational, industrial arts, adult education, and occupational educational programs. Along with its companion texts, it will provide an excellent foundation for the person who wants to make this trade a vocation.

ACKNOWLEDGMENTS

The authors wish to express their appreciation and acknowledge the following organizations for their assistance in the development of this text:

- The American Welding Society, for permission to use and adapt the Chart of Standard Welding Symbols.

- Hobart Brothers Company, Troy, Ohio, for comparison charts.

- Larry Jeffus, technical photographer and welding consultant.

- Lincoln Electric Company, Cleveland, Ohio.

- Miller Electric Manufacturing Company, Appleton, Wisconsin.

APPLICATION CHART — BASIC ARC WELDING

UNIT NUMBER	AUTO MECHANIC	BOILER MAKER	BRICKLAYER	CARPENTER	ELECTRICIAN	FARM EQUIPMENT REPAIR	GENERAL WELDING	IRON WORKER (ORNAMENTAL)	IRON WORKER (STRUCTURAL)	MACHINIST	PLUMBER	SHEET METAL WORKER	STEAMFITTER
1	■	■	■	■	■	■	■	■	■	■	■	■	■
2	■	■	■	■	■	■	■	■	■	■	■	■	■
3	■	■	■	■	■	■	■	■	■	■	■	■	■
4	■	■	■	■	■	■	■	■	■	■	■	■	■
5	■	■	■	■	■	■	■	■	■	■	■	■	■
6	■	■	■	■	■	■	■	■	■	■	■	■	■
7	■	■	■	■	■	■	■	■	■	■	■	■	■
8	■	■	■	■	■	■	■	■	■	■	■	■	■
9	■	■	■	■	■	■	■	■	■	■	■	■	■
10	■	■	■	■	■	■	■	■	■	■	■	■	■
11		■					■	■	■			■	■
12	■	■	■	■	■	■	■	■	■	■	■	■	■
13	■	■	■	■	■	■	■	■	■	■	■	■	■
14		■				■	■	■	■		■	■	■
15		■				■	■	■	■		■	■	■
16		■				■	■	■	■		■	■	■
17		■				■	■	■	■		■	■	■
18		■				■	■	■	■		■	■	■
19		■					■	■	■		■	■	■
20		■					■	■	■		■	■	■
21		■					■	■	■		■	■	■
22	■	■	■	■	■		■	■	■		■	■	■
23	■	■					■	■	■		■	■	■
24		■					■	■	■		■	■	■
25		■					■	■	■		■	■	■
26		■					■	■	■		■	■	■

Note: The above chart is in terms of suggested minimums only. The final choice of course content is a function of the individual instructor, often with the advice of an industry advisory committee.

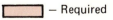 — Required

THE ARC WELDING PROCESS

One of the most important processes in industry is the fusion of metals by an electric arc. This is commonly called *arc welding* or *SMAW* (Shielded Metal Arc Welding).

Briefly, the process takes place in the following manner. The work to be welded is connected to one side of an electric circuit, and a metal electrode is connected to the other side. These two parts of the circuit are brought together and then separated slightly. The electric current jumps the gap and causes a continuous spark called an *arc.* The high temperature of this arc melts the metal to be welded, forming a molten puddle. The electrode also melts and adds metal to the puddle, figure 1-1.

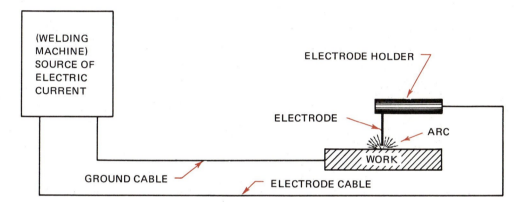

Fig. 1-1 Simple welding circuit

As the arc is moved, the metal solidifies. The metal fuses into one piece as it solidifies.

The melting action is controlled by changing the amount of electric current which flows across the arc and by changing the size of the electrode.

HAZARDS

Before arc welding is begun, the student should be fully aware of the personal dangers involved. Of course, the high-temperature arc and the hot metal can cause severe burns. However, the electric arc itself may be a hazard.

An electric arc gives off large amounts of ultraviolet and infrared rays. Infrared rays are also given off from the molten metal. Both types of rays given off from arc welding are invisible, just as they are when given off from the sun. They will cause sunburn, the same as they will from the sun, except that the rays given off from the electric arc, burn much more rapidly and deeply. Since these rays are produced very close to the operator, they can cause severe damage to the eyes in a very short period of time.

During arc welding there is a danger that small droplets of molten metal may leave the arc and fly in all directions. These so-called sparks range in temperature from 2,000 degrees F.

(1,093 degrees C.) to 3,000 degrees F. (1,649 degrees C.), and in size from very small to as large as 1/4 inch in diameter. They may cause burns plus they are a fire hazard when they fall on flammable material.

PROTECTIVE DEVICES

For protection from the rays of the arc and the flying sparks, the welding operator must use a helmet, figure 1-2, and other protective devices. The welding helmet is fitted with filter plates that screen out over 99% of the harmful rays. The helmet must be in place before attempting to do any arc welding. The arc is harmful up to a distance of 50 feet and all persons within this range must be careful that the rays do not reach their eyes.

Most welding helmets are made of pressed composition material or molded plastic. If they are dropped or if material is dropped on them they may be unfit for use. Each helmet is equipped with an adjustable headband. Any attempt to use wrenches or pliers to force the adjusting device may destroy the helmet.

The filter plate in each helmet is a special, costly glass which should be handled with great care. All filter plates should be protected from the flying globules of molten metal in the manner indicated in figure 1-3.

All welding stations should be equipped with curtains or other devices which keep the arc rays confined to the welding area. For the protection of others, the welder should make sure that these curtains are in place before starting any welding.

Fig. 1-2 A modern welding helmet

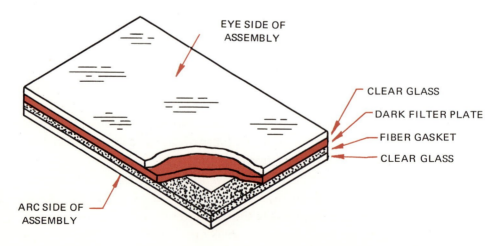

EYE SIDE OF ASSEMBLY

CLEAR GLASS

DARK FILTER PLATE

FIBER GASKET

CLEAR GLASS

ARC SIDE OF ASSEMBLY

Fig. 1-3 Filter plate protective assembly

Fig. 1-4 Note protective devices.

The arc rays will penetrate one thickness of cloth and cause sunburn. Therefore, the operator must protect himself with fire-resistant aprons, sleeves, and gloves to eliminate this hazard plus the hazard of fire, figure 1-4. In fact, all clothing worn by the operator should be reasonably flame resistant. Clothing which has a fuzzy surface can be a serious fire hazard, particularly if it is cotton.

Naturally, all other types of flammable material such as oil, wood, paper, and waste should be removed from the welding area before any welding is attempted. Each welder should be acquainted with the location and operating characteristics of all fire extinguishers.

REVIEW QUESTIONS

1. Why is clear glass or plastic used on the arc side of the filter plate assembly in figure 1-3?

2. What determines how fast the weld metal melts?

3. Who is responsible for the protection of workers in the area of the welding operation?

4. What type of rays are given off by the electric arc?

5. Can all of the light rays given off by the electric arc be seen?

SOURCES OF ELECTRICITY FOR WELDING

TYPES OF WELDING MACHINES

Electric current for the welding arc is generally provided by one of two methods. A transformer which reduces the line voltage can provide *alternating current* (AC). This current reverses direction 120 times per second. The transformer has no moving parts.

Direct current (DC) for the welding arc may be produced by a direct-current generator connected by a shaft to an AC motor. A gasoline engine or other type of power may also be used to turn the generator, figure 2-1. Direct current flows in the same direction at all times. In any case, the welding machine must have the ability to respond to the need for rapid changes in the welding voltage and current.

Fig. 2-1 An engine generator-type of DC welding machine

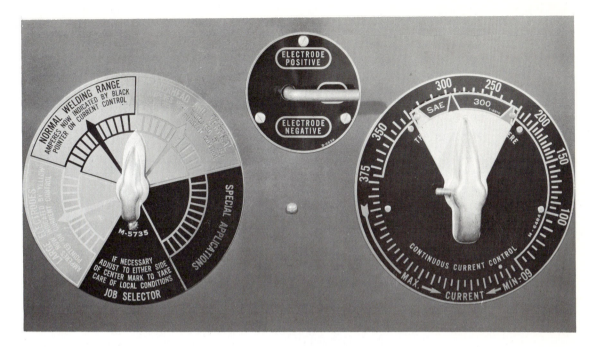

Fig. 2-2 Dual-current control

Both types of machines are widely used in industry, but the DC type is slightly more popular. Both are supplied in various sizes, depending on the use to which they are to be put, and are designated by the maximum continuous current in amperes which they can supply, (e.g., 150a or 300a).

The safe operation ampere rating of a welder is determined by duty cycle. (Duty cycle is noted on the nameplate of each welding machine along with the maximum amperage rating.)

The duty cycle of a welding machine is the percentage of a 10-minute period that the welder can operate at maximum output current setting. If a welder is rated at 300 amps. at a 60% duty cycle, the machine can be operated safely at 300 amps. 6 out of every 10 minutes. If the output amperage is set lower, then the duty cycle is increased.

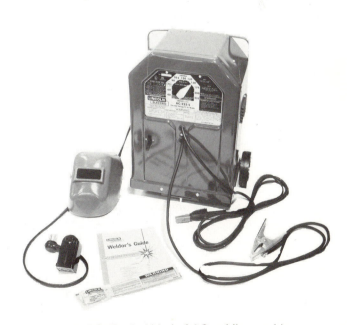

Fig. 2-3 Typical kind of AC welding machine

CURRENT CONTROL

Different types of welding operations require different amounts of current (amperes). Therefore, arc welding machines must have a way of changing the amount of current flowing to the arc.

The DC generator may have either a dual-current control, figure 2-2, or a single-current control. In the dual-current control type, two handwheels or knobs adjust the electrical circuits to provide the proper current to the arc. In the single-current type, a single wheel adjusts the current. The other controls on a DC machine are an on-off switch and a polarity switch.

In one type of AC welding machine, figure 2-3, the amount of current is selected by moving a knob to various heat settings.

The other popular type of AC machine has a movable core in the transformer. On this type of machine the operator selects the desired current rating by turning a handwheel.

Fig. 2-4 Combination AC-DC welding machine

Another type of machine which has gained popularity is the transformer-type with a built-in rectifier. The rectifier converts alternating current to direct current for welding. In some types of machines, the alternating current can be taken ahead of the rectifier when it is advantageous to use alternating current in the welding operation, figure 2-4.

It should be noted here that every electrode can be used on DC, but not on AC. DC current is more stable than AC current when welding in the vertical and overhead position. AC is best suited when using large, heavy-coated electrodes and when arc blow is not a problem. Arc blow is a magnetic force that tends to pull the arc from its normal path at times on DC current.

CARE AND PRECAUTIONS

- When a motor-generator type of welding machine is turned on, the operator should immediately check to see if the armature is rotating and that the direction of rotation is correct according to the arrow on the unit.

- Occasionally a fuse blows or a starter contact becomes burned or worn. Either of these conditions may cause the machine to overheat if it is left on. This can rapidly damage the welding machine.

- Starter boxes and fuse boxes carrying 220 or 440 volts should not be opened by the operator.

- The welding cable terminal lugs should be clean and securely fastened to the terminal posts of the machine. Loose or dirty electrical connections tend to overheat and cause damage to the terminal posts.

REFERENCE

Manufacturer's bulletin for the machine to be used.

REVIEW QUESTIONS

1. What is a direct-current welding circuit?

2. What does the term current control mean?

3. What effect do loose connections have on the welding circuit?

4. What are the three types of welding machines?

5. How many times per second does alternating current reverse direction?

Unit 3

THE WELDING CIRCUIT

PARTS OF THE WELDING CIRCUIT

In addition to the source of current, the welding circuit consists of:

- The work
- The welding cables
- The electrode holder
- The electrode

The *work* with which the welder is concerned may be steel plate, pipe, and structural shapes of varying sizes and thicknesses. It should be suitably positioned for the job being done. The work is a conductor of electricity and thus, is a part of the circuit.

The *welding cables,* figure 3-1, are flexible, rubber-covered copper cables of a large enough size to carry the necessary current to the work and to the electrode holder without overheating. The size of the cable depends on the capacity of the machine, and the distance from the work to the machine. A *ground clamp,* figure 3-2, is attached to the end of one of the cables, so that it may be connected to the work.

The *electrode holder,* figure 3-3, is a mechanical device on the end of the welding cable which clamps the welding rod or electrode in the desired position. It also provides

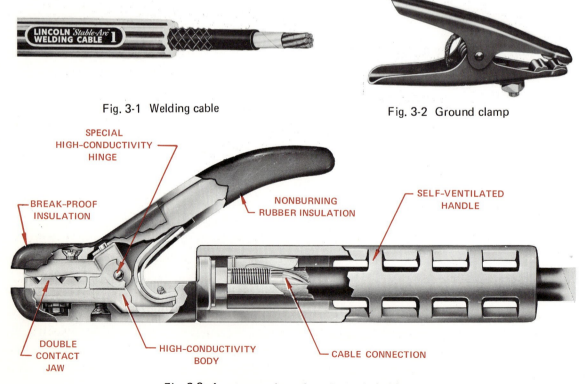

Fig. 3-1 Welding cable

Fig. 3-2 Ground clamp

SPECIAL
HIGH-CONDUCTIVITY
HINGE

BREAK-PROOF
INSULATION

NONBURNING
RUBBER INSULATION

SELF-VENTILATED
HANDLE

DOUBLE
CONTACT
JAW

HIGH-CONDUCTIVITY
BODY

CABLE CONNECTION

Fig. 3-3 A cutaway view of an electrode holder

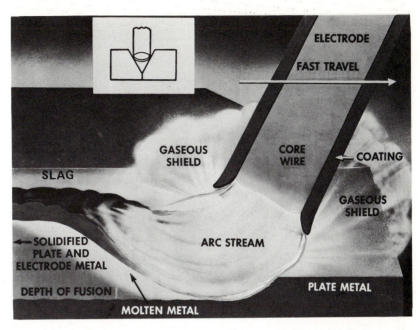

Fig. 3-4 Coated electrodes form a gaseous shield around the arc

an insulated handle, with which the operator can direct the electrode and arc. These holders come in various sizes depending on the amperage which they are required to carry to the electrode.

THE ELECTRODE

The electrode usually has a steel core. This core is covered with a coating containing several elements, some of which burn under the heat of the arc to form a gaseous shield around the arc, figure 3-4. This shield keeps the harmful oxygen and nitrogen in the atmosphere away from the welding area.

Other elements in the coating melt and form a protective slag over the finished weld. This slag promotes slower cooling and also protects the finished weld or bead from the atmosphere. Some coated electrodes are designed with alloying elements in the coating which change the chemical and physical characteristics of the deposited weld metal.

The result of using properly designed coated electrodes is a weld metal which has the same characteristics as the work, or base metal, being joined.

Electrodes are supplied commercially in a variety of lengths and diameters. In addition, they are supplied in a wide variety of coatings for specific job applications. These applications are discussed in other units.

The American Welding Society and The National Electrical Manufacturer's Association classify electrodes according to the type of coating, operating characteristics, and chemical composition of the weld metal produced. Chart 3-1 indicates the commonly used electrodes. There are many more. Most electrode manufacturers supply, free of charge, a chart of all the electrodes they make. Electrode manufacturers mark the grip end of the electrodes with the AWS classification number as indicated in chart 3-1.

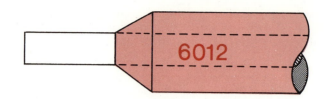

A W S NUMBER LOCATION

Chart 3-1

AWS Classification	Type	Current and Polarity	Tensile Strength P.S.I.	Yield Point P.S.I.	% Elongation in 2 inches	Applications
			MILD STEEL ELECTRODE CHART			
E-6010	Coated	d.c. Positive	65-72,000	53-58,000	27-30%	All-position, all-purpose, high-impact, high-ductility code welding
E-6011	Coated	a.c. or d.c. Positive	65-72,000	53-58,000	27-30%	All-position, primarily for producing welds with a.c. current equal to E-6010
E-6012	Coated	a.c. or d.c. Negative	68-78,000	58-68,000	20%	All-position, for poor "fitup" work and work where resistance to impact and low ductility are not too important
E-6013	Coated	a.c. or d.c. Negative	75,000	62,000	20%	For all-position work primarily with a.c. current. Compares to E-6012 series in general applications
E-6020	Coated	a.c. or d.c. Negative	68,000	56,000	32%	For flat-position welding, used in code welding. Being replaced slowly by E-6027 and E-7024
E-7014	Coated Iron Powder	a.c. or d.c. Negative	72-82,000	62-72,000	23-32%	High-speed production work, faster deposition rate than E-6012 or E-6013
E-7024	Coated Heavy Iron Powder	a.c. or d.c.	75-83,000	63-75,000	17-25%	Fast deposition rate, excellent appearance — for joints not requiring deep penetration
E-6027	Coated Iron Powder	a.c. or d.c. Negative	62-69,000	52-60,000	25-35%	Iron powder version of E-6020 — faster deposition for flat and horizontal positions

Note: Low-hydrogen electrode information may be found in unit 23.

Only the number is important when determining the type of electrode. The coating colors should not be depended on for recognition as they vary with manufacturers. Some producers also place dots and other markings on the coating. These are only trademarks, and are not to be confused with the numbers which appear only on the grip end of the electrodes.

Any of the electrodes shown in chart 3-1 may be purchased in a wide variety of sizes and lengths.

AWS CLASSIFICATION NUMBERS

All AWS (American Welding Society) numbers consist of three parts. For example, E-6010.

1. The E in all cases indicates an electric arc welding electrode or rod.

2. The number following the E (in this case, 60) indicates the minimum tensile strength of the weld metal in thousands of pounds per square inch (in this case, 60,000 p.s.i.). This number could be 80, 100, or 120 which would indicate minimum tensile strengths respectively of 80,000 p.s.i., 100,000 p.s.i., or 120,000 p.s.i.

3. In a four-digit number the third digit indicates the positions in which the electrode may be used: 1—indicates all positions; 2—flat or horizontal; 3—deep groove.

4. The fourth digit indicates the operating characteristics, such as polarity, type of coating, bead contour, etc.

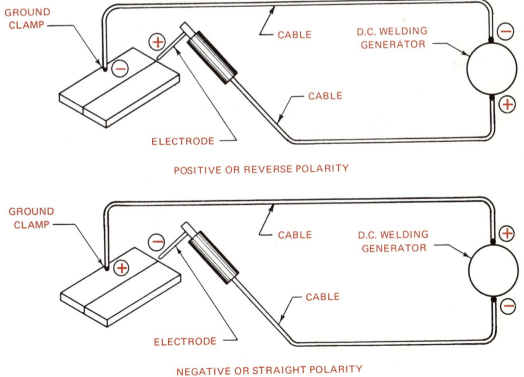

POSITIVE OR REVERSE POLARITY

NEGATIVE OR STRAIGHT POLARITY

Fig. 3-5 The welding circuit

POLARITY

In welding with direct current, the electrode must be connected to the correct terminal of the welding machine. This polarity may be changed by a switch on the welding machine. The polarity to be used is determined by the type of electrode and is indicated in the electrode chart, chart 3-1.

When the electrode is connected to the negative terminal (−), the polarity is called *negative or straight.* When connected to the positive terminal (+), it is called *positive or reverse,* figure 3-5. The use of incorrect polarity produces a poor weld. When welding with al-alternating current, polarity is not considered.

A simple test for checking the polarity of an electric welding machine is as follows:

1.   Place a carbon electrode in the electrode holder.

2.   Strike an arc. Maintain the puddle and weld for 5 or 6 inches.

3.   Check the plate for smears or black smudges. If these are present, the machine is in reverse polarity.

REFERENCE

Manufacturer's chart of electrodes.

REVIEW QUESTIONS

1.   What is the primary purpose of the AWS code markings on welding rods?

2.   What are the effects of oxides and nitrides in the weld metal?

3.   Referring to chart 3-1, what AWS type of electrode is used if the strength of the weld is the only characteristic of importance for a given job?

4.   If it is necessary to weld pressure pipe in the overhead position and the only machine available is an alternating-current type (AC), what AWS classification rod is used?

FUNDAMENTALS OF ARC WELDING

VARIABLES

Four things greatly affect the results obtained in electric arc welding. To make good welds, each one must be adjusted to fit the type of work done and the equipment being used.

They are:

- Current setting or amperage

- Length of arc or arc voltage

- Rate of travel

- Angle of the electrode

CURRENT SETTING

The current which the welding machine supplies to the arc must change with the size of the electrode being used. Large electrodes use more current than smaller sizes. A good general rule to follow is: when welding with standard coated electrodes, the current setting should be equal to the diameter of the electrode in thousandths of an inch.

Thus, a 1/8-inch electrode measures .125 inch and operates well at $125 \pm$ a few amperes. Similarly, a 5/32-inch rod measures .156 inch and operates well at $150 \pm$ a few amperes. The $\pm$ indicates that these electrodes will operate well in a range of current values either below or above the indicated amperage. For example, a value of $125 \pm$ 10 amperes indicates a range of values with a low of 115 amperes and a high of 135 amperes.

When indicating the diameter of the electrodes, reference is made only to the steel or alloy core of the rod, figure 4-1. The overall diameter including the rod coating is not the indicated electrode size.

Fig. 4-1   Measure core of rod

LENGTH OF ARC

The arc length is one of the most important considerations in arc welding. Variations in arc length produce varying results.

The arc length increases as the arc voltage increases. For example, an arc 3/16 inch long requires three times the voltage of a 1/16-inch arc, figure 4-2.

The general rule on arc length states: The arc length shall be slightly less than the diameter of the electrode being used.

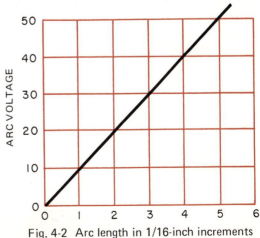

Fig. 4-2 Arc length in 1/16-inch increments

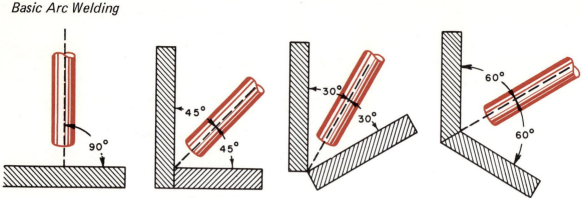

Fig. 4-3 Electrode splits angle of weld.

Thus, a 5/32-inch diameter electrode operates well between 1/8 inch and 5/32 inch of arc gap, or 20-22 arc volts according to the chart, figure 4-2.

It is almost impossible for the operator to measure the arc length accurately when welding. However, the welder can be guided by the sound of the arc. At the proper arc length, the sound is a sharp, energetic crackle. Proper arc length is determined by noting the difference in the sound of the arc when it is set too far, and at just about the right distance from the work. By practicing this, the operator will be able to judge good arc length by the distinctive sound.

RATE OF TRAVEL

The *rate of travel* of the arc changes with the thickness of the metal being welded, the amount of current, and the size and shape of the weld, or bead, desired.

The welding student should begin by making welds known as single-pass stringer beads. The arc length and arc travel should be such that the puddle of molten metal is about twice the diameter of the rod used.

ANGLE OF ELECTRODES

When welding on plates in a flat position, the electrode should make an angle of 90 degrees with the work. In other than flat work, good results are obtained if the rod splits whatever angle is being welded, figure 4-3. In general practice it is found that this angle may vary as much as 15 degrees in any direction without affecting the appearance and quality of the weld. The electrode angle should be no greater than 20 degrees toward the direction of travel.

REVIEW QUESTIONS

1. Using the general rule for current setting, what is the proper setting, to the nearest round figure for electrodes with the following diameters: 3/32 inch, 3/16 inch, and 1/4 inch?

2. From figure 4-2 and the general rule for arc length, what is the voltage across the arc when welding with a 3/16-inch electrode?

3. What is the arc voltage for a 1/8-inch diameter electrode?

4. If the first attempt at making a stringer bead produces a weld that is too narrow, what adjustment must be made in the rate of travel to produce a bead of the proper width?

5. What is the proper angle of the electrode in the following sketch?

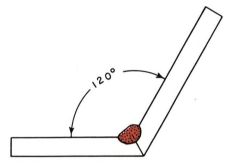

6. What indication does the operator have that the arc is the correct length for the diameter rod being used?

Unit 5

WELDING SYMBOLS

DESCRIBING WELDS ON DRAWINGS

Welding symbols form a shorthand method of conveying information from the drafts-man to the fabricator and welder. A few good symbols give more information than several paragraphs.

The American Welding Society has prepared a pamphlet, *Standard Welding Symbols* (AWS A2.0-68). This publication provides the draftsman with the exact procedures and standards to be followed so fabricators and welding operators are given all the information necessary to produce the correct weld.

The standard AWS symbols for arc and gas welding are shown in figure 5-1.

EXAMPLES OF THE USE OF SYMBOLS

Each of the symbols in this unit should be studied and compared with the drawing which shows its significance. They should also be compared with the symbols shown in figure 5-1.

In each of the succeeding units a symbol related to the particular job is shown together with its meaning. A study of these examples will clarify the meaning of the welding symbols.

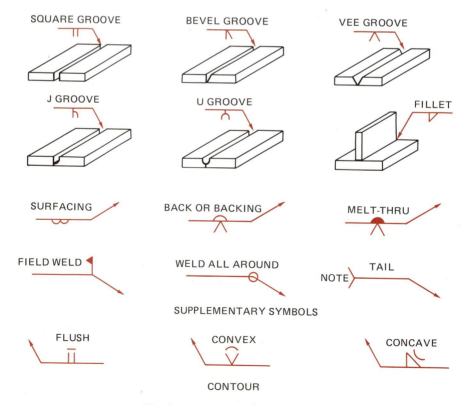

Fig. 5-1 Standard welding symbols

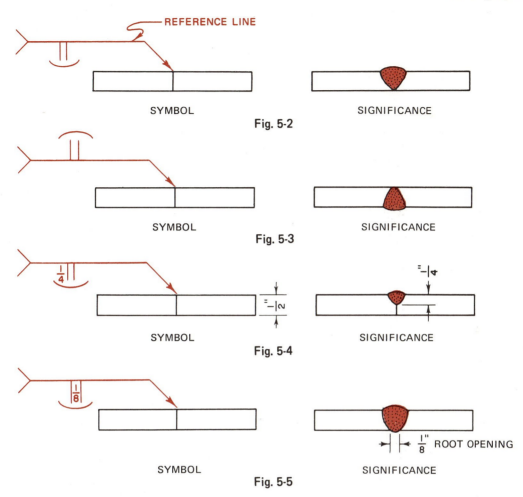

REFERENCE LINE

SYMBOL   SIGNIFICANCE

Fig. 5-2

SYMBOL   SIGNIFICANCE

Fig. 5-3

SYMBOL   SIGNIFICANCE

Fig. 5-4

SYMBOL   SIGNIFICANCE

$\frac{1}{8}$" ROOT OPENING

Fig. 5-5

The symbols from the chart are placed at the midpoint of a reference line. When the symbol is on the near side of the reference line, the weld should be made on the arrow side of the joint as in figure 5-2.

If the symbol is on the other side of the reference line, as in figure 5-3, the weld should be made on the far side of the joint or the side opposite the arrowhead.

All penetration and fusion should be complete unless otherwise indicated by a dimension positioned as shown by the 1/4 in figure 5-4.

To distinguish between root opening and depth of penetration, the amount of root opening for an open square butt joint is indicated by placing the dimension within the symbol, figure 5-5, instead of at one side of the symbol, as in the preceding drawing.

The included angle of beveled joints and root opening is indicated in figure 5-6, page 18. If no root opening is indicated on the symbol, it is assumed that the plates are butted tight unless the manufacturer has set up a standard for all butt joints.

The tail of the arrow on reference lines is often provided so that a draftsman may indicate a particular specification not otherwise shown by the symbol. Such specifications are usually prepared by individual manufacturers in booklet or looseleaf form for their engineering and fabricating departments. The specifications cover such items as the welding process to be used (i.e. arc or gas), the size and type of rod or electrode, and the preparation for welding, such as preheating.

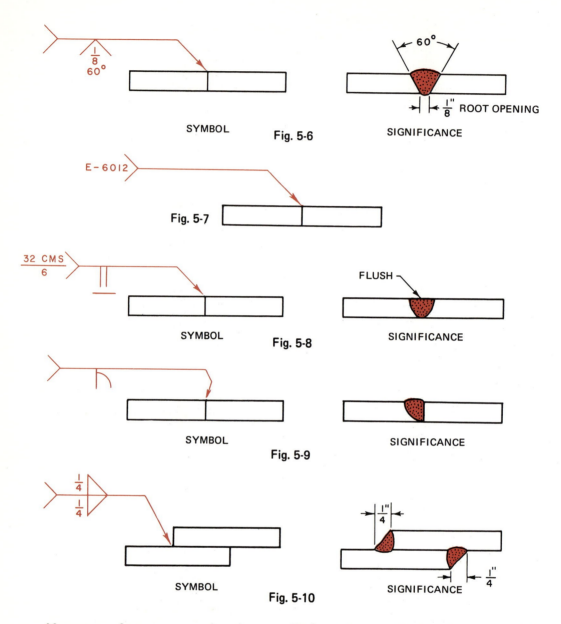

SYMBOL   **Fig. 5-6**   SIGNIFICANCE

Fig. 5-7

SYMBOL   **Fig. 5-8**   SIGNIFICANCE

SYMBOL   SIGNIFICANCE

Fig. 5-9

SYMBOL   SIGNIFICANCE

Fig. 5-10

Many manufacturers are using the new AWS publication *Welding Symbols* which gives very complete rules and examples for welding symbols, as well as a complete set of specifications with letters and numbers to indicate the process.

One method of indicating the type of rod to be used is shown in figure 5-7. This figure indicates that the butt weld is to be made with an AWS classification E-6012 electrode.

In figure 5-8, the rod to be used is indicated as a number 32 CMS (carbon mild steel) type and the 6 indicates the size of the rod in 32nds of an inch. In this case it is 3/16-inch diameter rod. In addition the symbol indicates that the finished weld is to be flat or flush with the surface of the parent metal. This may be accomplished by G = grinding, C = chipping, and M = machining.

When only one member of a joint is to be beveled, the arrow makes a definite break back toward the member to be beveled, figure 5-9.

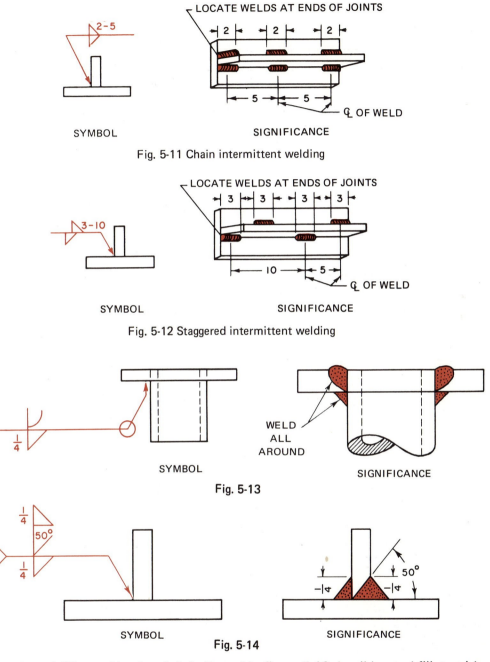

LOCATE WELDS AT ENDS OF JOINTS

SYMBOL   SIGNIFICANCE

Fig. 5-11 Chain intermittent welding

LOCATE WELDS AT ENDS OF JOINTS

SYMBOL   SIGNIFICANCE

Fig. 5-12 Staggered intermittent welding

WELD
ALL
AROUND

SYMBOL   SIGNIFICANCE

Fig. 5-13

SYMBOL   SIGNIFICANCE

Fig. 5-14

The size of fillet and lap beads is indicated in figure 5-10. In all lap and fillet welds, the two legs of the weld are equal unless otherwise specified.

If the welds are to be chain intermittent, the length of the welds and the center-to-center spacing is indicated, as in figure 5-11.

When the weld is to be staggered intermittent, the symbols and desired weld are made as in figure 5-12.

An indication that the joint is to be welded all around is shown by placing the weld all around symbol, as in figure 5-13.

Several symbols may be used together when necessary, figure 5-14.

Field welds (any welds not made in the shop) are indicated by placing the field weld symbol at the break in the reference line, as in figure 5-15.

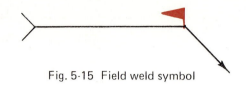

Fig. 5-15 Field weld symbol

REVIEW QUESTIONS

1. What is the symbol for a 60-degree closed butt weld on pipe?

2. What is the symbol for a U-groove weld with a 3/32-inch root opening?

3. What is the symbol for a double V, closed butt joint in plate?

4. What is the symbol for a 1/2-inch fillet weld in which a column base is welded to an H-beam all around?

5. What is the symbol for a J-groove weld on the opposite side of a plate joint?

STARTING AN ARC AND RUNNING STRINGER BEADS

The quality and appearance of an electric arc weld depend almost entirely on the following:

- Length of the arc
- Rate of travel
- Angle of the electrode
- Amount of current

Experimentation with each of these variables is helpful in learning correct welding procedures. This unit provides an opportunity to experiment with these variables and observe the results.

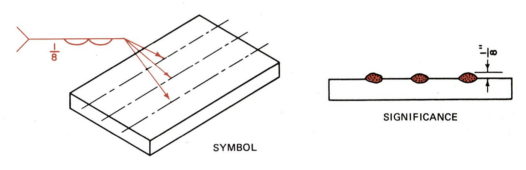

SYMBOL

SIGNIFICANCE

Fig. 6-1 Stringer beads

Materials

Steel plate 3/16″ or thicker 6 in. x 9 to 12 in.

DC or AC welding machine

1/8-inch or 5/32-inch diameter E-6012 or E-6013 electrodes

1. Start the machine, check the polarity and adjust the current setting as described in unit 4.

    CAUTION: Make sure all protective devices are in place. Use all recommended safety devices to protect the body, especially the eyes, from the arc rays. Failure to do so results in severe and painful radiation burns. Wear safety glasses.

2. Start the arc on the plate according to the method indicated in figure 6-2.

3. Listen for the sound indicating the correct arc length, and observe the behavior of the arc.

Fig. 6-2A The electrode is dragged along the plate.

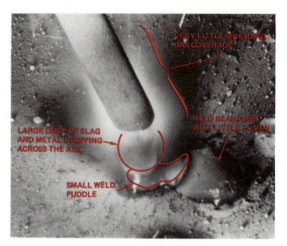

Fig. 6-2B. The electrode is raised
and the arc is established.

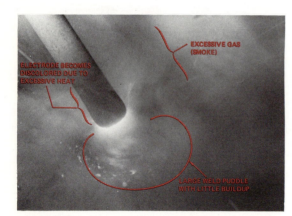

Fig. 6-2C. The electrode is held over the
starting spot until the puddle is established.

(Reprinted from Jeffus & Johnson, *Welding: Principles & Applications,* figures 11-2, 11-5, 11-6. © 1984 by Delmar Publishers Inc.)

4. Make straight beads or welds. Note that the electrode must be fed downward at a constant rate to keep the right arc length. Move the arc forward at a constant rate to form the bead.

Note: Right-handed welders will see better welding from left to right.
Left-handed welders should weld from right to left.

5. Remove the slag and examine the bead for uniformity of height and width.

CAUTION: When removing slag from a weld with a chipping hammer, eye protection is very important. Safety glasses should always be worn.

6. Continue to make stringer beads until each weld is smooth and uniform.

7. Make a series of beads similar to those in figure 6-3. Note the difference of each bead as the variables are changed.

A. NORMAL BEAD
B. ARC TOO LONG
C. ARC TOO SHORT
D. RATE OF TRAVEL TOO HIGH
E. RATE OF TRAVEL TOO LOW
F. CURRENT TOO LOW
G. CURRENT TOO HIGH
H. ROD ANGLE TOO LOW

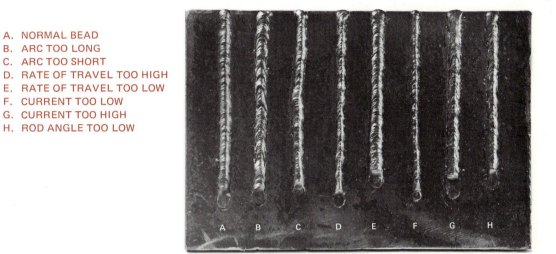

Fig. 6-3 Bead variables using E-6012 electrode and DC straight polarity

Chart 6-1

COMPARATIVE INDEX OF MILD STEEL AND ALLOY ELECTRODES									
AWS Classification	Air Products and Chemicals	Airco Welding Products	Alloy Rods Div.	Canadian Liquid Air Ltd.	Canadian Rockwell Ltd.	Hobart Bros. Company	Lincoln Electric	Murex Welding Products	Reid Avery Div.
E-6010	AP 6010 W	6010	AP-100 SW-610	LA 6011	R-60	10 60 AP	Fleetweld 5P	Speedex 610	Raco 6010
E-6011	AP 6011 W	6011 6011 C	SW-14	LA 6011 P	R-61	335 A	Fleetweld 35 35 LS 180	Type A 611 C 611 LV	Raco 6011
E-6012	AP 6012 W	6012	SW-612 PFA	LA 6012 P	R-62	12 12 A 212 A	Fleetweld 7	Type N 13 Genex M	Raco 6012
E-6013	AP 6013 W	6013 6013 D	SW-15	LA 6013 LA 6013 P	R-63	447 A 413	Fleetweld 37 57	Type U U 13	Raco 6013
E-7014	AP 7014 W	Easyarc 7014	SW-151.P.	LA 7014	R-74	14 A	Fleetweld 47	Speedex U	Raco 7014
E-6027	AP 6027 W	6027			R-627	27 H	Jetweld 2	Speedex 27	Raco 6027
E-7024	AP 7024 W	Easyarc 7024 7024 D	7024	LA 7024 LA 24 HD	R-724	24 24 H	Jetweld 1 3	Speedex 24 24 D	Raco 7024

REVIEW QUESTIONS

1. What must be controlled to make good arc welds?

2. What is a stringer bead?

3. What are the points to look for in a good weld?

4. What is the current value for a 5/32-inch diameter electrode?

5. What two factors best determine a correct arc?

RUNNING CONTINUOUS STRINGER BEADS

Running long stringer beads demands good control of the welding electrode if the beads are to be straight and uniform in appearance and size. Much practice is required to develop a high degree of skill.

Changing electrodes in the middle of the bead, or starting an arc which has been accidentally stopped is a basic and important skill. This unit provides experience in restarting a bead.

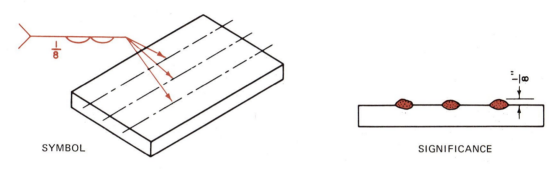

SYMBOL   SIGNIFICANCE

Fig. 7-1 Stringer beads

Materials

Steel plate, 3/16 inch thick, 6 in. x 9 to 12 in.

DC or AC welding machine

1/8- or 5/32-inch diameter E-6012 or E-6013 electrodes

Procedure

1. Start the machine, check the polarity, and adjust the current for the size of electrode being used.

2. Run a continuous stringer bead on the plate, using the full length of the electrode before stopping. Make this bead parallel to and about 1/2 inch from the edge of the plate, figure 7-2.

3. Run additional beads at 1/2-inch intervals, being sure to keep each bead straight. Check the arc length and rate of travel constantly to produce smooth, uniform beads.

4. Continue to make this type of weld until each one is of uniform appearance for its entire length.

5. Make a bead 2 or 3 inches long and stop the arc. Start the arc again ahead of the crater. Move the electrode back to the crater, using an extra long arc. Bring the rod down rapidly to the proper arc length and make sure that the new puddle just fuses into the last ripple of the crater. Proceed with the weld for another 2 or 3 inches and stop.

Fig. 7-2 Stringer bead on plate

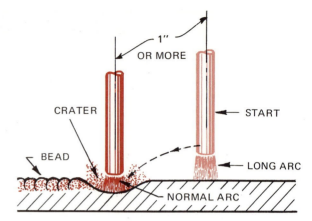

Fig. 7-3 Establishing a continuous bead

6. Continue this procedure until there is very little difference in appearance at the point where the arc was restarted. Figure 7-3 shows the right procedure.

7. Start an arc directly in the crater and notice the difference in the appearance of the connection, figure 7-4.

Note: The ideal length at which electrode stubs should be discarded is 1-1/2″.

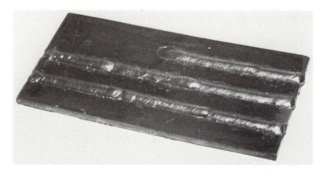

Fig. 7-4 Stopping and restarting stringer beads

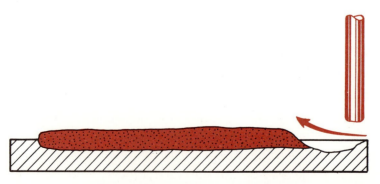

Fig. 7-5 Rod motion for filling crater

8. Try to eliminate the crater at the end of the finished bead by moving the arc back over the crater and finished bead slowly. As the arc is moved back gradually increase the arc length until the crater is filled as in figure 7-5.

9. Try to eliminate the crater by using a very short arc in the crater and pausing until there is enough buildup.

REVIEW QUESTIONS

1. What is the difficulty in trying to restart an arc directly in the crater?

2. What are the advantages of restarting a long arc ahead of the crater then backing up to the crater and shortening the arc?

3. Why is it necessary to gain skill in connecting the beads?

4. Why is it important for a welding student to learn to run a bead in a straight line?

5. What does the term polarity mean?

Unit 8

RUNNING WEAVE BEADS

It is often necessary when welding large joints or making cover passes to produce beads wider than stringer beads. These are called *weave beads*. Weaving is done with a back and forth sidewise motion of the electrode and a slow forward movement, figure 8-1.

NO MORE THAN 3 ELECTRODES WIDE

PAUSE HERE

Fig. 8-1 Straight weave

The height of the bead depends on the amount the electrode is advanced from one weave to the next. The number of ripples depends on the speed and frequency of the weaving motions.

The pause at each side of the weave is important for puddle flow and penetration. Failure to pause causes an undercut along the sides of the weld, figure 8-2.

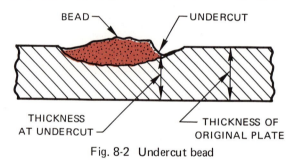

BEAD

UNDERCUT

THICKNESS AT UNDERCUT

THICKNESS OF ORIGINAL PLATE

Fig. 8-2 Undercut bead

Materials

Steel plate

1/8- or 5/32-inch diameter E-6012 or E-6013 electrodes

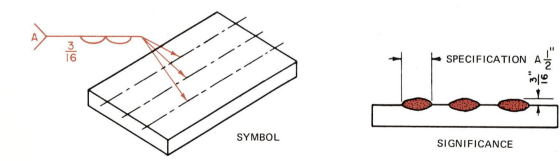

$\frac{3}{16}$

SYMBOL

SPECIFICATION A $\frac{1}{2}$"

$\frac{3}{16}$"

SIGNIFICANCE

Fig. 8-3 Weave beads

Procedure

1. Start the machine, check the polarity, and adjust the current for the size of the electrode being used.

2. Start an arc and make a bead approximately 3 times wider than the diameter of the electrode being used.

3. Continue to make weave beads until they have a uniform height and width for their entire length. See figure 8-4.

   Note: Beginners have a tendency to let the weld become progressively wider with each pass of the arc. In an attempt to correct this, there is a tendency to decrease each weave motion.

4. Try stopping the arc and fusing the new bead to the original. Be sure to *slag* (remove the slag with a wire brush or hammer) the crater before starting each bead. Practice this until the starting and stopping points blend in smoothly.

5. Change the length of pause at each side. Change the travel speed and the amount of advance.

6. Clean the beads and compare the results.

REVIEW QUESTIONS

1. Why should there be a definite pause at each side of the weave?

Fig. 8-4 Weave beads using the E-6012 electrode

2. What can be done to produce a weave bead with many fine ripples rather than a few coarse ones?

3. When restarting the bead, how does the rate of travel for the first two or three passes compare with the normal rate of travel?

4. What steps are taken to prevent undercutting?

5. How should the size of the molten puddle compare with the diameter of the electrode?

PADDING A PLATE

Experience in padding helps the welding student develop an eye for following a joint. It helps the student compare beads for uniform appearance. Padding is used for building up pieces prior to machining and for depositing hardfacing metal on construction equipment.

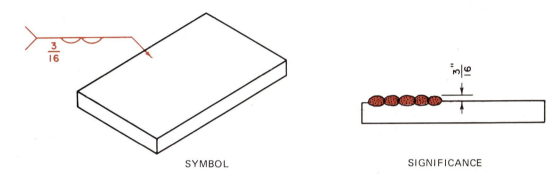

SYMBOL   SIGNIFICANCE

Fig. 9-1 The pad

Materials

2 pieces of plate 6 in. x 6 in.

1/8- or 5/32-inch E-6012 or E-6013 electrodes

Procedure

1. Establish an arc and run a stringer bead close to and parallel with the far edge of the practice plate.

   Note: Observe that a crater is left at the end of this weld. Prevent this crater in the following manner: Upon reaching the end of the plate, pull the electrode out of the crater, letting the heat die down. When the color has disappeared restart the arc in the crater, depositing a small amount of weld. This can be done several times to fill the crater.

2. Run more beads alongside of the previous bead, figure 9-2. Make sure that the far far edge of bead being deposited is in the center of the previous bead.

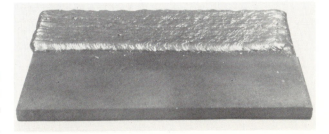

Fig. 9-2 A partially completed pad using E-6012 electrode (right-handed operator welding left to right)

      Note: The electrode must be directed at the point where the previous bead meets the base metal.

             Always chip the slag from each bead before welding.

3. Continue to cover the plate with weld using this technique. Keep the weld as straight as possible.

4. On another plate follow the same procedure and make a pad using the weave technique.

      Note: As the newly padded surface cools, the plate may bow upward. This can be corrected by welding a pad on the opposite side of the plate.

5. For additional practice, the plate can be turned 90 degrees for a second layer.

6. After the padding operation is finished, cut the material by saw or torch and examine the weld. There should be no holes or bits of slag imbedded in the weld.

REVIEW QUESTIONS

1. What is weld padding?

2. What characteristics of a weld are most easily examined through padding?

3. Where can the padding operation be used?

4. Why is it important to slag every bead before running the next one?

5. How is a washed-out area at the edge of a plate prevented?

SINGLE-PASS, CLOSED SQUARE BUTT JOINT

In making a single-pass, closed square butt joint, penetration is extremely important. Welding from one side not only makes complete penetration difficult, but the joint strength depends directly on the depth of penetration.

By experimenting with this type of joint and testing the results, one can determine the best procedures.

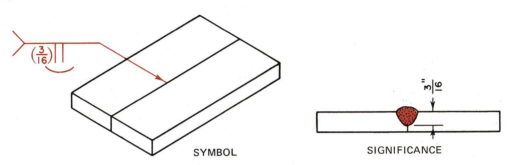

SYMBOL   SIGNIFICANCE

Fig. 10-1 Single-pass, closed square butt joint

Materials

Two steel plates, 1/4 inch thick, 2 in. x 9 in. each

1/8- or 5/32-inch diameter E-6012 or E-6013 electrodes

Procedure

1. Place the plates on the worktable so that the two 9-inch edges are in close contact.

2. Tack the two plates together using *tack welds* about 1/2 inch long. Start the tacks 1/2 inch to 1 inch from the ends of the plate to avoid having excessive metal and poor penetration at the start of the weld.

3. Proceed with the weld as in making stringer beads, but be very careful to keep the centerline of the arc exactly centered on the joint. Half the weld should be deposited on each plate. See figure 10-2.

4. Cool the finished assembly. Clean the bead and examine it for uniformity.

5. Check the depth of penetration of the weld by placing the assembly in a vise with the center of the weld slightly above and parallel to the jaws. Bend the plate toward the face of the weld so that the joint opens, figure 10-3. Examine the original plate edges.

6. Continue the bend until the weld breaks. Notice that the broken weld metal has a bright, shiny appearance, and that the metal that was not welded is much darker. This bright weld metal indicates the depth of penetration.

7. Make more joints of this type. Start with a setting of 150 amps and increase the amperage with each joint.

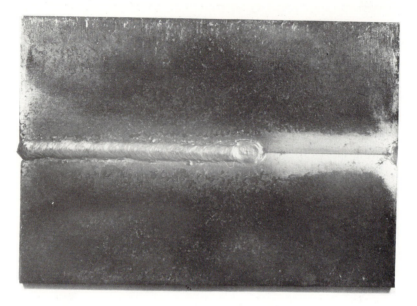

Fig. 10-2 Closed square groove butt joint

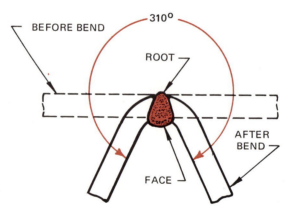

Fig. 10-3 Bend test for butt weld

8. Cool, break, and examine these test plates and compare the amount of penetration with that in the first weld made.

9. Set up another test plate and weld, using a weaving figure-8 motion, figure 10-4.

10. Cool, break, and examine this plate and compare the penetration with the other welds that have been made.

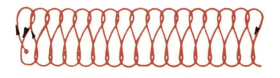

Fig. 10-4 Figure-8 weave

REVIEW QUESTIONS

1. How does increasing the amperage affect the depth of penetration?

2. How does the figure-8 weave bead affect the depth of penetration? Why?

3. Is there any advantage if the figure-8 weave is used in buildup operations?

4. What is penetration?

5. What is the purpose of a tack weld?

Unit 11

OPEN SQUARE BUTT JOINT

An open butt joint presents some additional problems in penetration. By changing the space between the plates, and by welding one set of plates from one side only and another set from both sides, it is possible to compare the quality of the welds and, particularly, the penetration. The open square butt joint differs from the closed butt joint in that the open-type joint has some spacing between the plate edges.

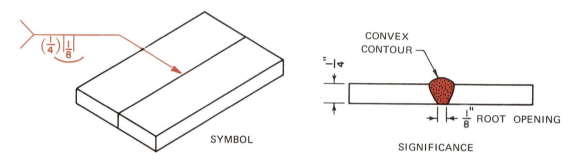

SYMBOL

CONVEX CONTOUR

$\frac{1}{8}$" ROOT OPENING

SIGNIFICANCE

Fig. 11-1 Open square butt joint

Materials

Two steel plates, 1/4 inch thick, 1-1/2 to 2 in. x 9 in. each

5/32-inch diameter E-6012 electrodes

Procedure

1. Place the two plates on the welding bench, align and space them, and tack them as shown in figure 11-2. Tacks should be long enough to withstand the strain of the expanding metal being welded without cracking.

2. Make the weld in the same manner as a closed butt joint, but use a slight amount of weaving to allow for the additional width of the joint.

3. Cool, clean, and inspect the finished weld for uniform appearance. Examine the root side of the weld for penetration.

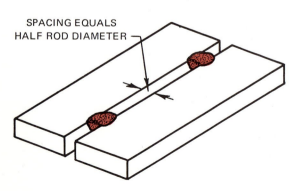

SPACING EQUALS
HALF ROD DIAMETER

Fig. 11-2 Setup for open butt joint

Fig. 11-3 Open square groove butt joint

4. Break the welded joint in the same manner as the closed butt joint. Check the amount of penetration. It should be a little more than one-half the thickness of the metal being welded.

5. Make additional open butt joints, using plate spacings narrower and wider than the first. After welding, check the plates for bead appearance and penetration.

   Note: If the spacing between the plates is too great it may be necessary to run another bead over the first one to build the weld to desired dimensions. The first bead is referred to as the burning-in or root-pass bead, and the second bead is called the finish bead.

6. Once beads of good appearance can be made consistently, make additional joints, but weld from both sides. Check these welds by cutting the plates in two and examining the cross section for holes and *slag inclusions* (nonmetallic particles trapped in the weld).

REVIEW QUESTIONS

1. What can be done to prevent holes or slag inclusions in the weld?

2. What advantages does the open butt joint have over the closed butt joint?

3. Sketch a cross section of the weld made in step 2 showing penetration and fusion.

4. Sketch a cross section of the weld made in step 5, with the spacing too wide. Show what is wrong with this weld.

5. What is the space between two pieces of plate being welded called?

Unit 12

SINGLE-PASS LAP JOINT

The welded lap joint has many applications and is economical to make since it requires very little preparation. For maximum strength it should be welded on both sides. A single pass or bead is enough for the plate used for this job. For heavier plate, several passes must be made.

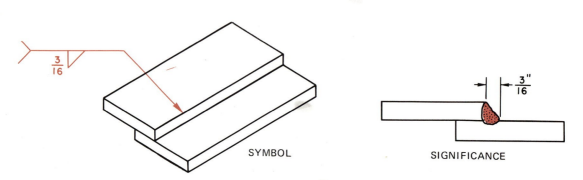

SYMBOL   SIGNIFICANCE

Fig. 12-1 Single-pass lap joint

Materials

Two steel plates, 3/16 inch thick, 2 in. x 9 in. to 12 in. each

1/8- or 5/32-inch diameter, E-6012 or E-6013 electrodes

Procedure

1. Set up the two plates as shown in figure 12-2. Make sure that the plates are reasonably clean and free of rust and oil. Be sure that the plates are flat and in close contact with each other.

2. Weld this joint with the electrode at an angle of 45 degrees from horizontal. Make sure that the weld metal penetrates the root of the joint, and that the weld metal builds up to the top of the lapping plate. Figure 12-3 shows the cross section, or end view, of a lap-welded joint, indicating the electrode angle and the size and

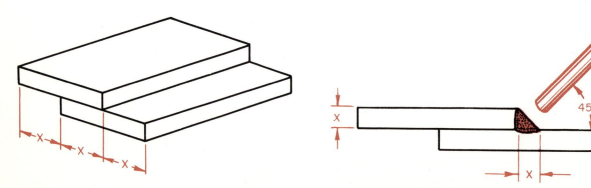

Fig. 12-2 Setup for lap joint

Fig. 12-3 Lap joint weld

Fig. 12-4 Properly welded lap joint

shape of the weld.   Notice that the weld makes a triangle with each side equal to the thickness of the plate (See "X" in figure 12-3.)

3. Cool and clean this bead and examine it for uniformity. Pay particular attention to the line of fusion with the top and bottom plates. This should be a straight line with the weld blending into the plate, figure 12-4.

   Note: Too slow a rate of travel deposits too much metal and causes the weld to roll over onto the bottom plate. This forms a sudden change in the shape. The extra metal is a waste of material, and actually weakens the joint by causing stresses, figure 12-5.

4. Continue to make this type of joint until beads of uniform appearance can be made each time. Be sure to weld both sides of the assembly.

5. Make a test plate in the same manner as the other lap joints but weld it on only one side.

6. Place this plate in a vise so that the top plate can be bent or peeled from the bottom plate, figure 12-6.   Bend this top plate until the joint breaks. Examine the break for penetration and uniformity. Another test may be made by sawing a lap-welded specimen in two and examining the cross section for penetration.

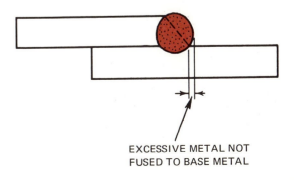

EXCESSIVE METAL NOT FUSED TO BASE METAL

Fig. 12-5 Lap joint with excessive weld metal

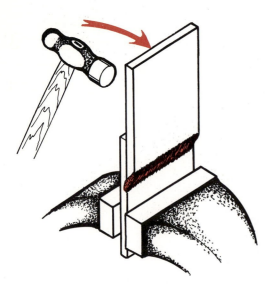

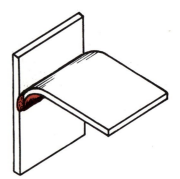

Fig. 12-6 Testing a lap weld

REVIEW QUESTIONS

1. Beginning students usually produce beads with an irregular line of fusion along the top edge of the overlapping plate. How is this corrected?

2. If the test shows lack of fusion at the root, how is this corrected?

3. What effect does too great a rod angle have on this type of joint?

4. What factor is most important in determining the location of the bead on a lap weld?

5. What is the result of depositing too much weld metal in a lap weld?

SINGLE-PASS FILLET WELD

The fillet or T-weld is similar to the lap weld but the heat distribution is different. It has many industrial applications.

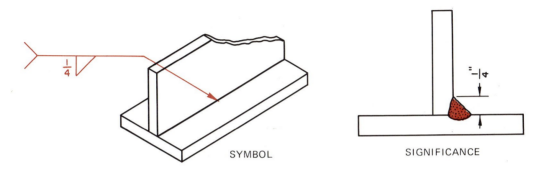

Fig. 13-1 Single-pass fillet weld

Materials

Two steel plates, 3/16 inch or 1/4 inch thick, 3 in. x 9 to 12 in. each

1/8- or 5/32-inch diameter, E-6012 electrodes

Procedure

1. Set up the two plates, figure 13-1. Be sure that the tack welds used to hold the plates in place are strong enough to resist cracking during welding, but not large enough to affect the appearance of the finished weld. This can be done by using a higher amperage and a higher rate of arc travel when tacking.

2. Make the fillet weld in much the same manner as the lap weld was made. The electrode angle is essentially the same. Both legs of the 45-degree triangle made by the weld must be equal to the thickness of the work for the full length of the joint, figure 13-2.

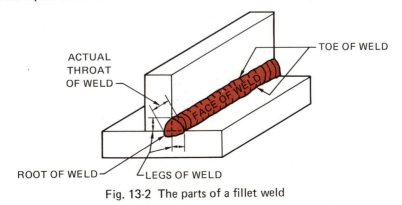

Fig. 13-2 The parts of a fillet weld

3. Clean each completed weld and examine the surface for appearance. Look for poor fusion along both edges of the weld. Examine it for undercutting on the up-standing leg. If undercutting does exist, it is probably being caused by either too long an arc or too high a rate of travel.

4. When good fillet welds can be made every time, make a test weld on only one side of the joint. Then bend the top plate against the joint until it breaks, figure 13-3. Examine the break for root penetration and uniform fusion.

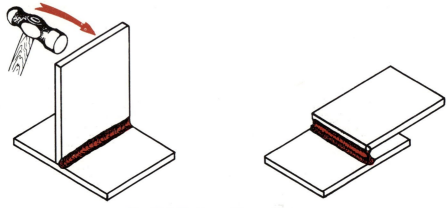

Fig. 13-3 Testing a fillet weld

REVIEW QUESTIONS

1. What can be done to prevent a hump at the start of the weld?

2. What causes undercutting?

3. What can be done to correct poor penetration and fusion in the root of the weld?

4. How is the size of a fillet weld measured?

5. If a fillet weld appears to be more on the flat plate than on the vertical plate how can it be corrected?

MULTIPLE-PASS LAP JOINT

When heavier plate is used it is impossible to cover the joint with one bead. Joints of this type require three or more passes or beads. The additional beads require that the rod angle be changed for each bead.

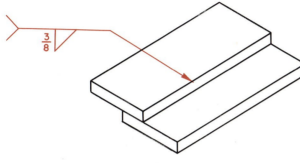

SYMBOL

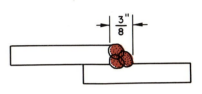

SIGNIFICANCE

Fig. 14-1 Multiple-pass lap joint

Materials

Two steel plates, 3/8 inch thick, 2 in. x 9 to 12 in. each

5/32-inch diameter E-6012 electrodes

Procedure

1. Set up the plates as for a single-pass lap joint.

2. Make the weld in three passes, being sure to clean each bead before going on to the next.

3. Check the angle of the electrode as each bead is deposited. Adjust this angle to suit the bead being welded. Each bead must be deposited at a different angle. Figure 14-2 indicates the rod angle and position of the arc for each bead.

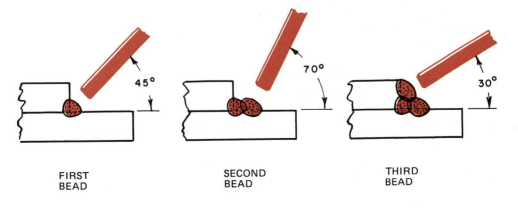

FIRST BEAD

SECOND BEAD

THIRD BEAD

Fig. 14-2 Multiple-pass weld

Fig. 14-3 Multiple-pass lap weld

Note: Observe that the second bead covers all but a very small portion of the first bead. This is necessary if the finished joint is to have a 45-degree angle face. The most common fault of beginners is not covering the bead enough, and then trying to stretch the third bead to correct the original fault. This results in a finished weld of low quality.

4. Continue to make this type of joint until beads of acceptable appearance are made each time. Check a test plate by cutting it in two and examining the section for penetration and slag inclusions.

REVIEW QUESTIONS

1. Figure 14-2 indicates a different electrode angle for each pass. Why is this necessary?

2. If the finished bead has a tendency to flatten out and become too wide, what steps are taken to insure that the face of the weld is at an angle of 45 degrees?

3. Is it necessary to slag the weld between beads? Why?

4. How wide should one bead be for a multiple-pass lap weld?

5. How much of the second bead should be visible when the third bead has been applied?

MULTIPLE-PASS FILLET WELD

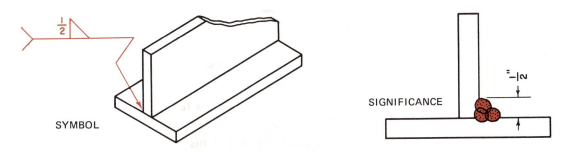

SYMBOL

SIGNIFICANCE

Fig. 15-1 Multiple-pass fillet weld

The heavy plate used in this unit requires 3 beads to complete the fillet.

Materials

Two steel plates, 1/2 inch thick, 2 in. x 9 to 12 in. each

5/32-inch diameter, E-6012 electrodes

Procedure

1. Set up and tack the plates in the manner used for the single-pass fillet welds.

2. Make a three-pass fillet weld, following the procedures shown in figures 15-1 and 15-2. Pay close attention to the angle of the electrode. When making the third pass, pass, check the arc length frequently to make sure that an undercut does not develop along the upstanding leg of the weld.

3. Make additional fillet welds, welding both sides of the joint. Do not make all beads on one side of the joint first, but rather alternate the sequence.

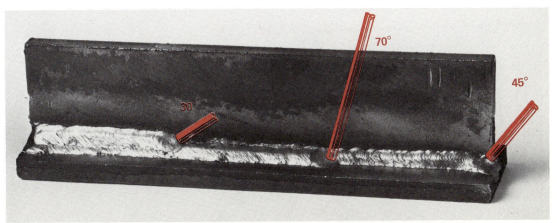

Fig. 15-2 Three-pass fillet weld using E-6012 electrode (right-handed operator)

4. Compare the distortion made by alternating sides with that made when only one side is welded.

5. Check the finished test plates in the same manner as the single-pass fillet welds were checked.

REVIEW QUESTIONS

1. What is the effect if the first bead on one side is followed by all three beads on the opposite side before completing the other two beads on the initial side? Why?

2. What should the height of the bead be for a fillet weld on 1/2-inch plate?

3. This unit indicates a weld made with three beads in two layers. How is a third layer applied?

4. How does the temperature of the base metal affect the overall appearance of the weld?

5. How does a hole in the first bead affect the second and third beads?

WEAVING A LAP WELD

To lap-weld heavy plate with a single pass, a weaving motion of the rod is necessary. This requires a special technique to avoid an irregular, defective weld.

The weaving motion produces an oval pool of molten metal. The weld makes an angle of about 15 degrees with the edge of the plate.

The angle of the puddle varies with the amount of current, length of arc, and speed of welding as well as with the thickness of the welded metal. In general, larger welds at higher amperages need an angle slightly greater than 15 degrees.

● DENOTES PAUSE IN ARC MOTION

Fig. 16-1 Weaving a lap weld in heavy plate

Materials

Two steel plates, 1/4 inch thick, 2 in. x 9 to 12 in. each

5/32-inch diameter, E-6012 electrodes

Procedure

1. Position the plates for a lap weld.

2. Weld the joint using a weaving motion to keep the bottom edge of the molten puddle ahead of the top of the puddle.

3. When making this type of weld, pause with the arc at the top of each weave motion but not at the bottom of each weave, figure 16-1. This method prevents burning away or undercutting of the top of the weld.

4. Cool, clean and examine the face of the finished weld. Look for undercutting of the top leg and overlapping of the bottom leg of the weld.

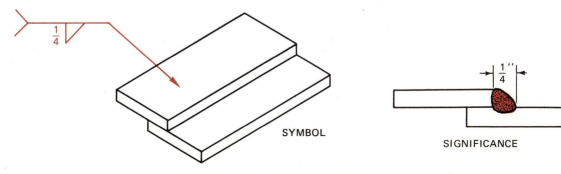

SYMBOL

SIGNIFICANCE

Fig. 16-2 Weaving a lap weld

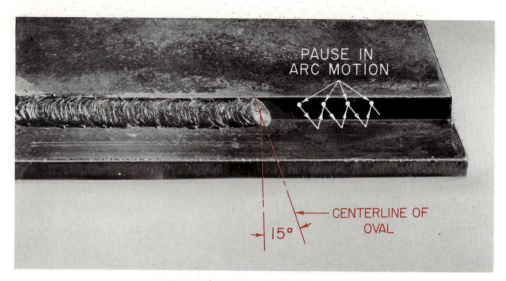

Fig. 16-3 Weaving a lap weld

5. Break this test plate and examine the root of the weld for good penetration.

6. Make another lap weld with the centerline of the molten puddle at a 90-degree angle to the line of the weld. Compare the line of fusion of the bottom leg of the weld with that of the previous bead.

7. Make more welds and examine them for appearance. All differences in ripple shape and spacing are caused by differences in arc control.

REVIEW QUESTIONS

1. Other than pausing at the top of each weave, how can irregularities and undercutting be controlled?

2. Does the undercutting referred to in this unit have the same appearance as the under-cutting of a fillet weld?

3. Step 6 indicates an electrode motion with no lead at the bottom of the weave cycle. How does this affect the appearance of the finished bead and the line of fusion with the bottom plate?

4. Make a sketch of the cross section of the weld made in step 6. Show the correct shape for a lap weld with a dotted line.

5. How does the time for weaving a weld compare with that for a multiple-pass weld?

Unit 17

WEAVING A FILLET WELD

This unit provides more practice in weaving beads to produce a multiple-pass weld of large size.

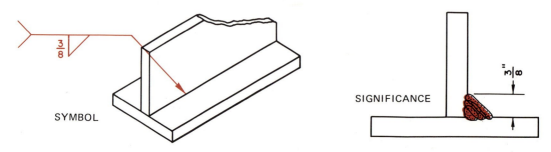

SYMBOL

SIGNIFICANCE

Fig. 17-1 Weaving a fillet weld

Materials

Two steel plates 3/8 inch thick, 2 in. x 9 to 12 in. each

5/32-inch diameter E-6012 electrodes

Procedure

1. Set up the plates as in figure 17-1.

2. Weld the joint using the electrode angle and weave motion described in unit 16. Weld a 3/8-inch fillet (i.e., each leg of the triangle formed by the weld should measure 3/8 inch).

   Note: The difference between a weave lap weld and a weave fillet weld is that the pause at the top of the puddle must be slightly longer for the fillet weld. Also, the length of the arc striking the up-standing leg must be kept very short to prevent undercutting.

3. Continue to make this type of weld. Examine each bead as it is made to determine what corrections are necessary to produce welds with uniform ripples and fusion.

4. Weave another bead over the original bead, using a very short arc and a definite pause as the arc is brought against the upstanding leg. Then bring the arc toward the bottom plate at a normal speed so that the bottom of the puddle leads the top of the puddle by 15 to 20 degrees. Return the arc to the top of the fillet rapidly with a rotary motion, figure 17-2.

5. Clean and inspect this bead for undercutting along the top leg of the weld and for poor fusion along the bottom edge. Also check for uniformity of the bead ripples.

6. Weave a third bead over the first two so that the bottom of the bead leads the top slightly more than 20 degrees. As the arc reaches the bottom of the fillet, try hooking the bead by moving the arc along the bottom before returning to the top of the weld.

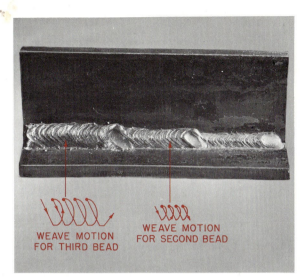

Fig. 17-2 Weave motions for a fillet weld

7. Clean and inspect weld as in step 5.

REVIEW QUESTIONS

1. What effect does hooking the bead have on the finished bead?

2. In making the second and third beads, is welding accomplished on the return stroke of the weave?

3. A common mistake in welding this type of fillet is depositing too much metal on the bottom plate. How is this fault corrected?

4. Is the bead made at step 6 good on 3/8-inch plate? Why?

5. Is undercutting more or less of a problem on the T joint than it is on the lap joint? Why?

Unit 18

BEVELED BUTT WELD

This important joint can be very strong if it is well made. The multiple-pass procedure used in this unit is likely to produce a better joint than a thick single-pass method.

The beveled butt joint requires skill in weaving beads of two different widths, cleaning the preceding bead, and controlling the width of the bead.

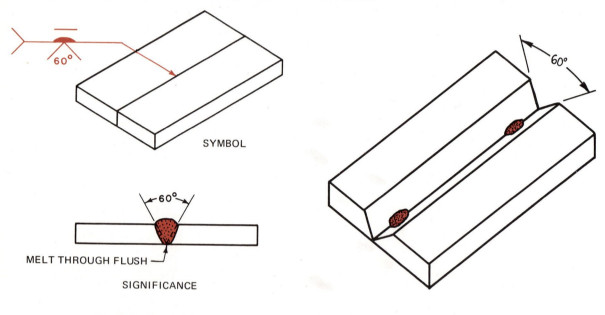

SYMBOL

MELT THROUGH FLUSH

SIGNIFICANCE

Fig. 18-1 Beveled butt weld

Fig. 18-2 Setup for beveled butt weld

Materials

Two steel plates, 3/8 inch thick, 4 in. x 9 to 12 in. each with one long edge beveled at 30 degrees

5/32-inch diameter E-6012 electrodes

Procedure

1. Align the plates on the welding bench and tack them as shown in figure 18-2.

2. Run a single-pass stringer bead in the root of the V formed by the two plates.

3. Clean the first bead and deposit a second bead by using a slight weaving motion. Allow the arc to sweep up the sides of the bevel in order to give this bead a slightly concave surface.

4. Clean the bead and run the third bead using a wide weaving motion. Do not allow the weld to become too wide. The actual width of the face of the weld should be slightly wider than the distance between the top edges of the V. Figure 18-3 is a cross-section view showing the size and contour of each bead.

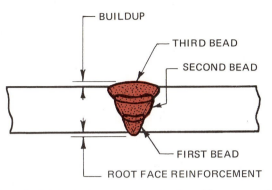

Fig. 18-3 Multiple-pass beveled butt weld, cross-section

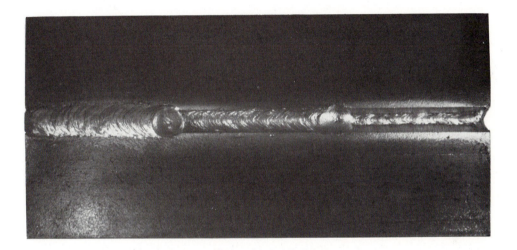

Fig. 18-4 Multiple-pass beveled butt weld

5. Clean and examine the finished weld for root penetration, evenness of fusion lines, and equal spacing of the ripples. Any variation in fusion, penetration, or ripple is caused by variations in arc manipulation. Uniform results can only be obtained by following uniform procedure. See figure 18-4.

    Note: Usually only three beads are required to make a bevel butt weld in 3/8-inch plate. However, if the first two beads are thin, do not attempt to make up for this by building the third bead much heavier. Instead, apply a normal third bead and a fourth if necessary. Heavy or thick buildups in one pass tend to produce holes and slag inclusions.

6. Align and tack a second set of plates as before, but leave an opening at the root of the V about one-half the rod diameter in size.

7. Weld these plates in the same manner as the first set and inspect visually.

REVIEW QUESTIONS

1. Why should the second bead have a slightly concave surface?

2. What effect does leaving a slight gap at the root of the V have on the finished bead?

3. If it is difficult to make the root pass with the plates gapped because of burn-through, what step is taken to correct this difficulty? Why?

4. How should the width of the finished bead compare with the width of the joint?

OUTSIDE CORNER WELD

Outside corner welds are frequently used as finished corners after they have been smoothed by grinding or other means, figure 19-1. In this case, the shape of the bead and the smoothness of the ripples are very important. Roughness, caused by too much or not enough weld metal and uneven ripples, requires a lot of smoothing. This results in higher cost.

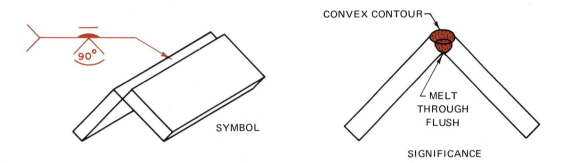

Fig. 19-1 Outside corner fillet

Materials

Two steel plates, 3/8 inch thick, 2 in. x 9 to 12 in. each

5/32-inch diameter E-6012 electrodes

Procedure

1. Set up the plates and tack them as shown in figure 19-2.

2. Make the weld in three or more passes as for a beveled butt joint. When making the final pass, observe all the precautions for weave welding to avoid any possibility of the finished bead overhanging the plate edges, figure 19-4. Place the assembly on the welding bench so that the weld can be made in the flat position.

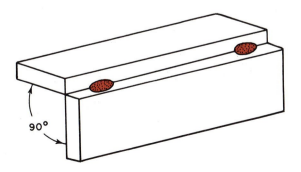

Fig. 19-2 Setup for outside corner weld

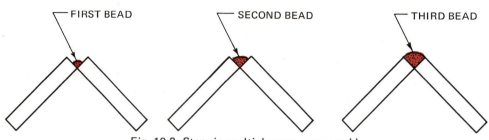

Fig. 19-3 Steps in multiple-pass corner weld

3. Check the finished weld for uniform appearance and to see if the angle of the plates is 90 degrees. Make any necessary corrections when setting up the next set of plates.

4. Test the weld by placing the assembly on an anvil and hammering it flat. Examine the root fusion and penetration, figure 19-5.

5. Make additional joints of this type, checking each weld for uniform appearance and for the shape shown in figure 19-4.

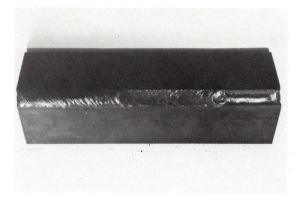

Fig. 19-4 Multiple-pass outside corner weld

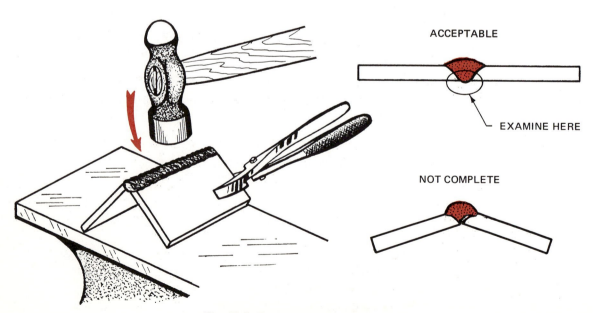

Fig. 19-5 Testing a corner weld

REVIEW QUESTIONS

1. How are poor fusion and uneven penetration corrected?

2. How are deep holes between the ripples in a finished weld corrected?

3. What can be done to prevent the final pass from building up too much, causing a hump along the line of fusion?

4. Is undercutting possible on an outside corner weld? How?

Unit 20

FLAT OPEN CORNER AND FILLET WELDS (IRON-POWDERED ELECTRODE)

This unit provides an opportunity to gain skill and knowledge in the use of heavy-coated electrodes. These electrodes are recommended for use in the downhand or flat position only. This unit also provides experience in adjusting the electrode angle, arc length, and current setting.

Research and development have produced an electrode of this type with a large amount of iron powder added to the coating. This type of electrode makes a weld with good physical characteristics, and very good appearance and contour. The slag is so easily removed that it is described as self-cleaning. It should also be noted that 30% to 50% of the filler metal of a given iron-powdered electrode comes from the coating.

The current values and arc voltages required to make this weld are quite high. As a result, the weld metal deposits rapidly. As much as 17 pounds of weld metal per hour may be deposited when using an iron-powdered electrode.

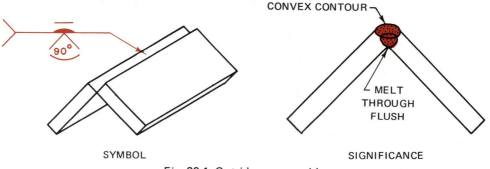

SYMBOL SIGNIFICANCE

Fig. 20-1 Outside corner weld

Materials

Steel plate, 3/8 inch thick, 2 in. x 9 to 12 in.

5/32-inch diameter x 14-inch E-7024 electrode

Procedure

1. Set up and tack the plates as indicated.

2. Adjust the current according to the manufacturer's recommendations, or use the formula for current settings. For an iron-powdered electrode 30% should be added to the current value, so that the minimum setting is about 200 amperes.

3. Make a root pass with little or no weaving motion.

4. Clean and examine this bead for appearance and note the ease of slag removal. Also examine the end of the electrode and note that the metal core has melted back into the coating to form a deep cup.

   Note: The cupping characteristic makes this rod good for *contact welding.* In contact welding the coating is allowed to touch the work. In this manner the operator does not have to maintain the arc length. However, this limits the possibility of controlling

Fig. 20-2 Outside corner weld with heavy-coated electrode

the electrode, so the width of the weld is limited to stringer beads. For this reason most experienced operators prefer to maintain a free arc which can be controlled.

5. Apply a second bead with a weaving motion and a rate of travel that builds up a bead as thick as the plates being joined.

6. Clean and inspect the bead. Especially notice the line of fusion, appearance of ripples, and the shape of the finished joint. Figure 20-2 shows both beads.

7. Make a series of joints of this type, but increase the current for each joint by 10 to 15 amperes until a final joint is made at 250 amperes. Also vary the amount of *lead* (the angle at which the electrode points back toward the finished bead).

8. Clean and inspect the joints. Hammer each joint flat on an anvil. Examine the weld metal for holes and slag inclusions, and compare the grain sizes.

   Note: Valuable practice in making fillet welds with the E-7024 electrode is gained by using the plates welded into outside corner welds, steps 1 through 7.

9. Set up the assembly to weld the inside corner as shown in figure 20-3. Make sure that any slag, which may have penetrated this inside corner from the original welds, is thoroughly cleaned from the joint.

10. Adjust the current values as in step 2, and proceed to make a single-pass 3/8-inch fillet weld. The electrode should point back toward the finished weld at an angle of 15 to 20 degrees.

11. Clean and inspect this bead for the shape of the weld, straight, even fusion line between the weld and plate, and any evidence of holes or slag inclusions in the

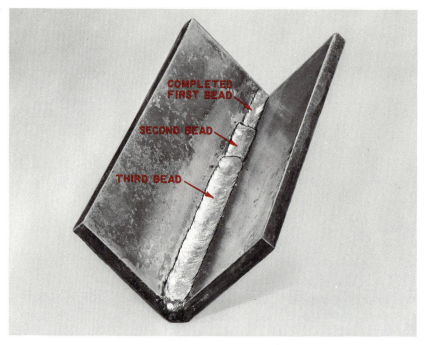

Fig. 20-3 Fillet weld with heavy-coated electrode

finished bead. Slag inclusions are usually caused by either poor cleaning of the original assembly, or an electrode angle that caused the slag to flow ahead of the arc and molten pool.

12. Run a second and third bead on top of the first, using a uniform, slightly weaving motion to produce enough width. Weld other joints with current values up to 250 amperes.

13. Clean and inspect, as in step 11.

14. Set up plates and make a three-pass fillet weld as in steps 10 through 12, figure 20-3.

15. Break this weld by hammering the two plates together. Check for penetration and grain structure.

16. Weld more joints with one leg of the final assembly flat and the other vertical. Try varying the electrode angle from more than 45 degrees to less than 45 degrees from horizontal, and check the results for fusion and bead shape.

17. Make additional joints as in steps 10 through 12, but incline the plates so that the welding proceeds in a slightly downhill direction. Then make some joints with the weld proceeding in a slightly uphill direction.

18. Clean and inspect these joints and compare the results in each case with the welds made when the joint was in a perfectly flat position.

REVIEW QUESTIONS

1. What polarity does the manufacturer recommend for this electrode?

2. It is suggested that various rod angles be tried while making these joints. What does this experiment show?

3. Too much current applied to a standard-coated electrode causes the coating to break down and burn some distance up the electrode. Is this true with the iron-powdered type of electrode?

4. How do the higher amperages used in step 7 affect the surface appearance of the joint?

5. How does the speed of making these fillet welds compare with the speed of similar welds made with an E-6012 electrode?

6. What can be done to correct the tendency of the slag to run ahead of the arc and make holes in the bead?

7. How does the cleaning time for welds made in this unit compare with that for similar welds made with E-6010 and E-6012 electrodes?

Unit 21

DEEP-GROOVE WELD
(IRON-POWDERED ELECTRODE)

Although deep groove welds can be made with a wide variety of electrode types, this unit provides practice and experience in using an electrode designed specifically for this type of joint.

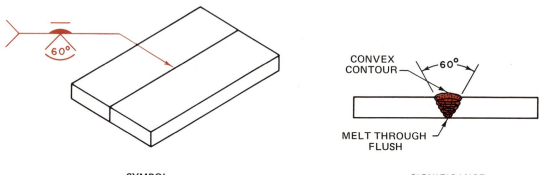

SYMBOL

SIGNIFICANCE

Fig. 21-1 Deep groove weld

Materials

Steel plates, 3/4 inch thick, 3 in. x 9 to 12 in. each

5/32-inch diameter, E-6027 electrodes

Procedure

1. Align and tack the beveled plates as shown.

2. Adjust the current for heavy-coated electrode and make the root pass. Use a free arc rather than the contact method.

3. Clean and inspect this bead, especially for shape, and observe the ease of slag removal.

4. Apply additional beads in layers 1/8 to 3/16 inch thick until the joint is completed. Use a weaving motion as necessary to provide a good surface appearance.

   Note: This electrode is designed to cause the weld metal to wash up the sides of the groove and form a bead with a slightly concave surface from which the slag can be easily removed between passes. Also notice the absence of undercutting with this electrode. Figure 21-2 shows a joint made in a series of steps to show the bead shape.

5. Clean and inspect the finished job for fusion at the root of the joint as well as along the edges of the weld. Cut a cross section from the finished joint and inspect for holes and slag inclusions. If the equipment is available, break the test section and inspect the grain structure.

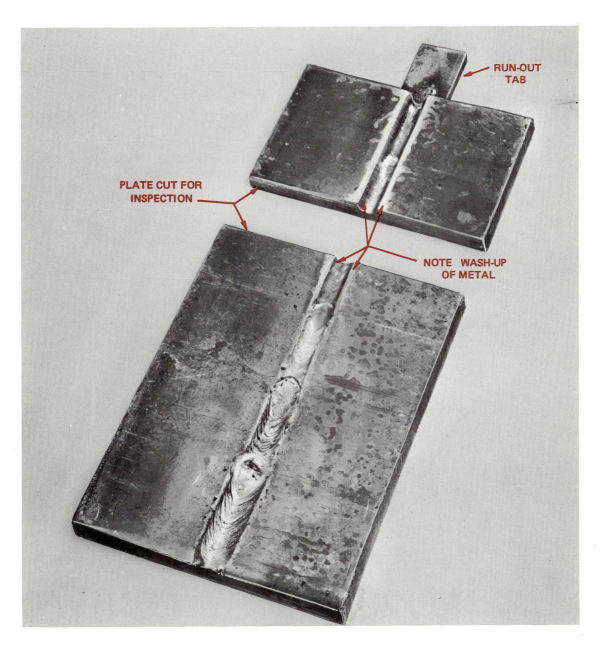

RUN-OUT
TAB

PLATE CUT FOR
INSPECTION

NOTE WASH-UP
OF METAL

Fig. 21-2 Deep groove weld with E-6027 electrode

Figure 21-2 shows the plates provided with a *run-out tab.* This is a device used to provide for continuing the weld beyond the ends of the work. It is then cut or broken from the work. It eliminates the necessity of filling the crater at the extreme edge of the job. This tab can be applied to any type of joint which ends suddenly. It is time-saving when the weld must have the same shape and size for its entire length. Run-out tabs are especially helpful in some automatic welding processes, in which there is no opportunity to manipulate the arc to fill the crater.

REVIEW QUESTIONS

1. How does slag removal for this type electrode compare with that for previous welds?

2. What are the advantages of heavy-coated electrodes?

3. What surface contour should intermediate passes have on deep-groove welds?

4. Can the E-6027 electrode be used for weave beads?

5. Why is a run-out tab used on some joints?

REVERSE-POLARITY WELDING

Direct current straight-polarity type electrodes are applied by a straight-line, steady motion. These electrodes produce a smooth, crowned bead when properly applied.

This unit covers the basic knowledge needed for the correct manipulation of E-6010 and E-6011 type arc welding electrodes. The E-6010 and E-6011 electrodes must be applied with a definite whipping motion. This makes a rougher surface appearance, but the penetration and bead shape is uniform in all positions.

DOWNHAND BEAD WITH WHIPPING MOTION AND HOT CRATER IN THE FLAT POSITION

Materials

Two steel plates, 3/8 inch thick, 6 in. x 6 in. each

5/32-inch diameter, E-6010 or E-6011 electrodes

Procedure

1. Set up a 3/8-in. thick, 6 in. x 6 in. plate in the flat position.

2. Set the amperage at 130 to 150 for the 5/32-inch electrode.

3. The welder should have good visibility of the arc area.

4. Hold the electrode perpendicular and pointed 10 to 20 degrees in the direction of travel.

1 TO 2 $-\frac{1}{4}$ IN. MOTION

2 TO 3 $-\frac{3}{16}$ IN. MOTION

SLIGHT HESITATION AT 1 AND 3

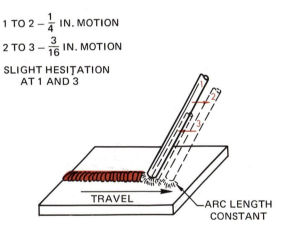

TRAVEL

ARC LENGTH CONSTANT

Fig. 22-1 Hot-crater method

5. Strike the arc and maintain standard arc length using a whipping motion.

   Note: The whipping motion shown by figure 22-1 is a forward and backward motion in the direction of travel. This motion may vary in length, but could be 1/4-inch motion ahead and 3/16-inch back. At the end of the back motion, a slight pause deposits the bead. The forward motion controls the amount of penetration. Various bead results can be attained by varying the pause, travel motion length, and speed of travel.

6. Practice this whipping motion in the flat position until a uniform, closely rippled bead is produced.

BEAD WITH WHIPPING MOTION AND COOL CRATER IN THE FLAT POSITION

Materials

Two steel plates, 3/8 inch thick, 6 in. x 6 in. each

5/32-inch diameter, E-6010 or E-6011 electrodes

Procedure

1. Set up a 3/8-in. thick, 6 in. x 6 in. plate in the flat position.

2. Set the amperage, position, and angles as for a downhand bead.

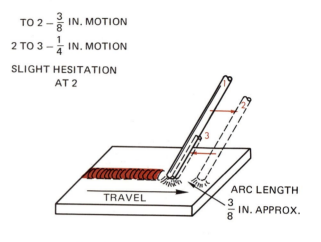

TO $2 - \frac{3}{8}$ IN. MOTION

2 TO $3 - \frac{1}{4}$ IN. MOTION

SLIGHT HESITATION AT 2

TRAVEL

ARC LENGTH $\frac{3}{8}$ IN. APPROX.

Fig. 22-2 Cool-crater method

3. Strike an arc and, using a whipping motion, proceed as in figure 22-2.

   Note: When welding *out of position* (any position other than flat) a slight change in the whipping action is needed. To overcome the pull of gravity on a molten mass, a cooler puddle is desirable. This is done by pausing at the end of the forward motion, and holding a long arc. On the backward motion, a normal arc length is used with no pause in the puddle.

4. Practice this until uniform results are attained. See figure 22-3.

   Note: The fillet welding technique for E-6010 and E-6011 electrodes is similar to that for other electrodes. The first bead, however, is a downhand bead.

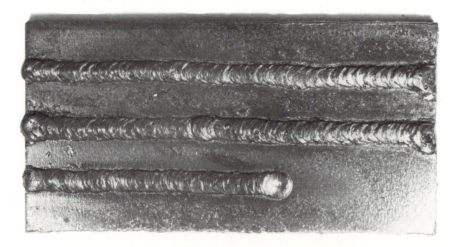

Fig. 22-3 Stringer beads using E-6011 electrode

REVIEW QUESTIONS

1. What is the difference in the way the E-6010 and E-6012 electrodes are handled?

2. When is the hot-crater technique used in welding with the E-6010 electrode?

3. What advantages does the cool-crater method have over the hot-crater method in the vertical-up welding position?

4. Is the electrode positive or negative in reverse-polarity welding?

5. What positions can the E-6010 and the E-6011 electrodes be used in?

Unit 23

LOW-HYDROGEN ELECTRODES

The low-hydrogen type of electrode replaces the E-6010 and E-6012 electrodes in many industrial and construction applications. The welding student should have knowledge of the characteristics and uses of this electrode.

The low-hydrogen electrodes are required in the welding of construction equipment, shipbuilding and nuclear power plants. The name, low hydrogen, refers to the fact that the coatings are free of the elements that contain hydrogen (moisture). By eliminating hydrogen, difficult steels can be welded with little or no preheat.

A low-hydrogen electrode has a core of mild steel or low-alloy steel. The most popular low-hydrogen electrode is the E-7018 type which has a coating of at least 30% iron powder. The arc is moderately penetrating and the slag is easily removed. These electrodes are recommended for the welding of alloy steels, high-carbon steels, malleable iron, spring steels and cast steels.

Welding may be done in all positions with electrode sizes up to 5/32 inch. Larger diameters are used for fillet and groove welds in the horizontal and flat positions. The mechanical properties produced by low-hydrogen electrodes are far superior to those found in conventional electrodes such as the E-6010 and E-6012. Tensile strength of 120,000 p.s.i. is possible in the as-welded condition.

CHARACTERISTICS AND TYPES

The gaseous shield formed by the electrode coating uses carbon dioxide as a shielding element. The gaseous shield formed by low-hydrogen electrodes should contain no hydrogen.

There are several types of low-hydrogen electrodes and it is important that the correct one be used, chart 23-1. For student training the E-7018 electrode is recommended.

Chart 23-1

					LOW-HYDROGEN ELECTRODE COMPARISON CHART					
AWS Classification	Air Products and Chemicals	Airco Welding Products	Alloy Rods Div.	Canadian Liquid Air Ltd.	Canadian Rockwell Ltd.	Hobart Bros. Company	Lincoln Electric	Murex Welding Products	Reid Avery Div.
E-7016	AP 7016 W	7016 7016 M	70 LA-2		Tensilarc 76			HTS HTS 18 HTS 180	Raco 7016
E-7018	AP 7018 W	Easyarc 7018 MR 7018 AC Codearc 7018 MR	Atom-Arc 7018 7018-1 SW-47	Superarc 18 LA 7018 Atom-Arc 7018 LA 7018 B	Hyloarc 78	718 718 LMP	Jetweld LH-70 Jet-LH 78 Jet-LH 75	Speedex HTS MR HTS M MR 718	Raco 7018
E-7028		Easyarc 7028		LA 7028 LA 7028 B	Hyloarc 728	728	Jetweld LH 3800	Speedex 28	Raco 7028

REASONS FOR USING LOW-HYDROGEN ELECTRODES

1. To provide better physical and mechanical properties.
2. To reduce bead and under-bead cracking in certain kinds of steel.
3. To improve *ductility* (to make the metal easier to work).
4. To reduce the temperature needed to preheat the weld.
5. To provide better low-temperature impact properties.

PROCEDURE GUIDELINES

- E-XX15 electrodes are used with DC reverse polarity. E-XX16 and E-XX18 are used with AC or DC reverse polarity. These electrodes are all-position type electrodes.

- E-XX28 are iron-powder electrodes. They are used in the flat position, with AC or DC reverse polarity.

- Do not use a whipping motion with low-hydrogen electrodes.

- The arc must be held very close.

- Stringer beads should be used rather than weave beads in all positions. The weave bead is likely to trap slag and gas in the bead.

- When making multiple-pass beads, all slag must be removed from each pass.

- Never restart a used electrode. It will cause porosity.

- Pinholes, usually found at the start or the end of a weld, may be caused by:
    1. Incorrect arc striking or stopping
    2. Moisture in the electrode coating
    3. Chipped coating on the electrode
    4. Moisture in the weld
    5. Arc length too great

- When striking an arc, always strike ahead of the point where the weld is to begin. Shorten the arc immediately, pushing the weld metal back to the starting point.

- When stopping the arc at the end of the joint, keep a short arc. If washout occurs, chip the slag, clean, and restrike as before.

- Low-hydrogen electrode coatings pick up moisture easily. Therefore, they should be stored in a thermostatically controlled oven or dry box. A temperature of 300 to 400 degrees F. is required.

- The operator should always examine every electrode he places in his electrode holder. A chip in the coating of a low-hydrogen electrode makes a hard spot and pinholes in the weld. A chipped rod can also cause the arc to be unstable.

- In some structural applications, the base metal may contain moisture. Preheating the weld area immediately before welding insures successful results.

When welding out of position, undercutting may occur at the edges of the weld. To overcome this hold a closer arc and, in the case of vertical-up welding, use a slight inverted U-shaped weave and hesitate on each side. Better control can be obtained by using an electrode one size smaller. The welding current should not be too high.

Arc blow (the tendency of the arc to wander off its path) is sometimes a problem in out-of-position welding with low-hydrogen electrodes. This is a greater problem with large electrodes (5/32- to 1/4-inch size). To overcome this:

1. Insure a good ground.
2. Clean the base metal as well as possible.
3. Keep the arc short.
4. Insure correct amperage.

REVIEW QUESTIONS

1. How is a low-hydrogen electrode different from others?

2. When welding with a low-hydrogen electrode, what is the most important thing for the operator to keep in mind?

3. Why is moisture bad for low-hydrogen electrodes and base material?

4. What is the tensile strength of a bead made with E-7018 electrodes?

5. Why is it difficult to restart an arc with a used E-7018 electrode?

HORIZONTAL WELDING

When the ability to make good welds in the flat position has been developed, it is important that out-of-position welding be learned. The easiest out-of-position welding is horizontal welding. It is different from flat welding because of the effect of gravity on the molten puddle.

Many jobs must be welded in the horizontal position. This is because the welded part cannot be moved, due to its size or location.

WELDING A HORIZONTAL PAD

Materials

Steel plate 1/4-inch or heavier, 4 inches x 8 inches.

1/8-inch E-6010 and E-7018 electrodes with a DC welding machine or 1/8-inch E-6011 electrode with an AC welding machine.

Procedure

1. Set the material up with the surface in a vertical position.

2. Set the current at 110-125 amps for 1/8-inch E-6010 electrode.

3. Right-handed operators weld from left to right.

   Note: Hold the electrode parallel with the floor no more than 10 degrees below horizontal. The electrode should also be angled toward the direction of travel 15 to 20 degrees.

   The E-6010 and E-6011 electrodes are applied with a whipping motion, but low-hydrogen electrodes are not.

4. Start the arc and make the whipping motion.

   Note: The molten metal will have a tendency to drop down to the bottom of the puddle if the electrode is not handled correctly. To overcome this sag, change the amount of pause at the back of the puddle and shorten the arc length.

   The low-hydrogen electrode is applied with a very short arc along a straight line.

   The same angles are used for both electrodes.

5. Continue making the horizontal pad by welding each bead above the preceding bead, figures 24-1 and 24-2.

   Note: The ability to do out-of-position welding requires considerable practice.

Forward progress is approximately 1/16" with each motion. Motion is straight back and forth with the pause just right of the crater center.

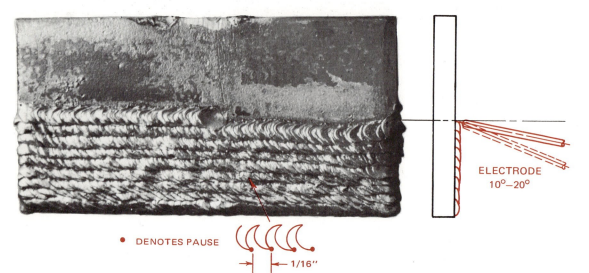

• DENOTES PAUSE

ELECTRODE 10°–20°

← 1/16" →

Fig. 24-1 Horizontal pad using E-6010 electrode (right-handed operator travelling left to right)

HORIZONTAL V-GROOVE BUTT JOINT

Materials

Two pieces of plate 1/4-inch or heavier, 2 inches x 8 inches, with a 30-degree bevel on one edge of each

1/8-inch E-6010 and E-7018 electrodes

DC welding machine

Procedure

1. Tack weld the material and position it for a horizontal weld. It may be necessary to experiment with the root opening to insure 100% penetration of the weld.

No motion with this electrode. Hold very close arc length.

10° MAXIMUM

Fig. 24-2 Horizontal pad using E-7018 electrode

Motion is straight back and forth on all stringer beads, with slight hesitation in crater. Progress is approximately 1/16'' with each motion.

Fig. 24-3 Horizontal V-groove butt joint using E-6010 electrode

2. Weld the root pass with a stringer bead.

   Note: Make sure the slag is removed from all passes. Run the beads in the order shown in figures 24-3 and 24-4.

3. The number of beads for a joint may vary, but they are always applied from the bottom up.

4. The surface appearance should be smooth, with no undercutting or overlapping.

5. When a good weld can be made in this manner, try welding heavier plate using 5/32-inch electrodes.

6. The V-groove should be tested by grinding the root side and face side flush with the surface. Cut 1-inch strips across the weld and test by bending the welded area in a vise. The cut will show any pinholes in the weld.

   CAUTION: Personal safety precautions must be observed in out-of-position welding. Falling spatter and sparks are always present.

No motion is required. Use close arc length.

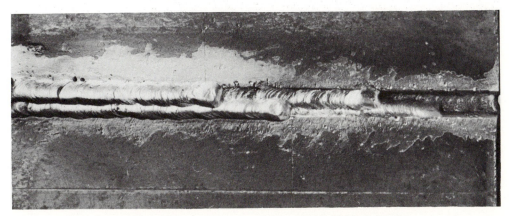

Fig. 24-4 Horizontal V-groove butt joint using E-7018 electrode

REVIEW QUESTIONS

1. What is horizontal welding?

2. What additional factor has to be considered in controlling the size and shape of the bead in horizontal welding?

3. What does whipping mean in welding and why is it used?

4. Is whipping necessary with the low-hydrogen electrode?

5. Is it practical to make weave beads in the horizontal position? Why?

OVERHEAD WELDING

When the ability to do horizontal welding is developed, the overhead fillet weld should not be difficult. The procedure is very similar. It is often necessary to use the overhead position in industry.

As in all welding processes, the welder must be comfortable to do a good job. Overhead welding can be tiring when it is done for a long period of time. Anything the operator can do to help keep the hand steady is an advantage.

OVERHEAD FILLET WELD

Materials

1/4-inch or heavier plate, 2 inches x 8 inches

1/8-inch E-6010 and E-7018 electrodes

DC welding machine

Procedure

1. Tack weld two pieces of plate together for a T joint.
2. Place the material in the overhead position. A holding tab can be tacked to the top of the plate.

   Note: For best results the joint line should be only slightly above eye level.

3. The root pass should be centered in the joint. A slight whipping motion is used with E-6010. E-7018 is not used with a whipping motion.
4. If it is difficult to start the arc, hold a longer arc and hesitate more.
5. The order in which the beads are run with each type of electrode is shown in figures 25-1 and 25-2.

First pass motion straight back and forth with slight hesitation in crater. Second pass oscillated with pauses at dot (●).

Fig. 25-1 Two-pass overhead fillet weld using E-6010 electrode (right-handed welder travelling left to right)

No motion required. Use close arc length.

Fig. 25-2 Three-pass overhead fillet weld using E-7018 electrode
(left-handed operator)

6. The electrode should be at an angle of about 45 degrees with the surface of the metal. It should point in the direction of travel at an angle of about 10 degrees to 20 degrees.

   Note: The electrode angle can be changed to force the weld metal to go where it should.

7. The second pass of the E-6010 electrode is also applied with a whipping motion as shown in figure 25-1.

   Note: There must be a pause at the top and back of the crater on each weave. If there is a buildup of weld metal on the vertical plate, the amount of pause and length of the arc must be adjusted.

8. Set up another pair of plates and make an overhead fillet weld using the stringer pass method with 1/8-inch E-7018. No whipping motion is required. Apply the beads in the order shown in figure 25-2.

REVIEW QUESTIONS

1. Why is the overhead welding position more difficult than the horizontal position?

2. What should be done if there is trouble in starting the arc?

3. What is the difference in the way the E-7018 and E-6010 electrodes are used on this joint?

4. Why is welding in the overhead position more dangerous than welding in other positions?

5. What should be the difference between a weld made in the overhead position and a weld made in the flat position?

Unit 26

VERTICAL WELDING

Probably the most difficult welding position for the beginner is the vertical position. The welding student should plan to spend considerable time practicing the various applications of this position. As in any out-of-position welding, safety equipment and clothing are very important.

Vertical welding can be done by welding uphill or by welding downhill. Welding vertical down is performed on sheet metal through 3/16-inch thickness and on some piping applications. Welding vertical up is performed on plate and structural parts 1/4 inch and heavier.

In the construction field, vertical welding ability is necessary because most weldments are so large they can't be positioned so the weld would be in the flat position.

As in the preceding units, 1/8-inch E-6010 and 1/8-inch E-7018 electrodes are used here. Pictures are used to show the student what the different beads should look like. The electrode motion, where applicable, is shown on the pictures.

VERTICAL-UP BEADS

Materials

DC welding machine

1/4-inch or heavier plate

1/8-inch E-6010 and 1/8-inch E-7018 electrodes

Procedure Using E-6010 Electrodes

1. Position the plate in the positioner with the surface vertical at eye level.

2. At approximately 110 amps, start the arc at the bottom of the plate. The electrode angle should be straight out from the plate and pointing upward from 5 to 10 degrees.

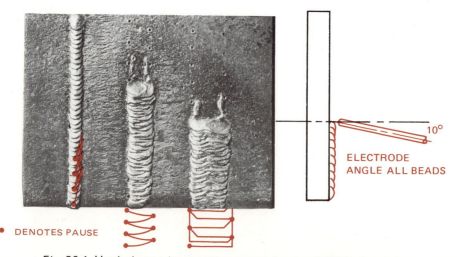

10°

ELECTRODE
ANGLE ALL BEADS

● DENOTES PAUSE

Fig. 26-1 Vertical-up stringer and weave beads using E-6010 electrodes

First pass stringer, no motion, straight up. Second pass inverted slightly U-shaped weave with hesitation at sides. Third pass straight side-to-side weave with pause at each side.

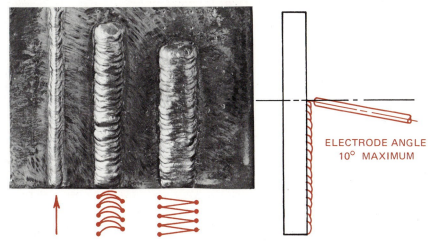

ELECTRODE ANGLE
10° MAXIMUM

Fig. 26-2 Vertical-up stringer and weave beads using E-7018 electrode

> *Note:* The stringer bead is applied with a straight up long arc movement and a straight down short arc. The weaves are applied with a side-to-side motion.

3. Always chip and wire brush the slag from each bead.

4. Practice each type of bead separately, by padding a plate.

Procedure Using E-7018, Low-hydrogen Electrodes

1. Position the plate in the positioner with the surface vertical at eye level.

2. At approximately 125 amps, start the arc at the bottom of the plate. Keep a short arc for the E-7018 electrode.

3. Stringer beads are applied with no motion other than the smooth forward progress of the bead.

4. The weave beads are applied with a slight inverted U-shaped side-to-side motion.

5. Practice each type of bead by padding a plate.

VERTICAL-UP MULTI-PASS FILLET WELD

Materials

DC welding machine

1/4-inch or heavier plate

1/8-inch E-6010 and 1/8-inch E-7018 electrodes

Procedure

1. Tack weld pieces of plate to form a T joint.

2. Apply the stringer bead and two weave beads, as shown in figures 26-3 and 26-4.

Weaves have a slight pause at the sides to prevent undercut.

First Pass

Second Pass Weave

Third Pass Weave

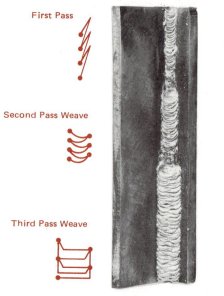

Fig. 26-3 Three-pass vertical-up fillet weld using E-6010 electrode

Hold a very close arc.

First Pass, Straight-up

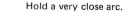

Second Pass, Inverted Slightly U-shaped Weave

Third Pass, Side-To-Side Weave With Pause At Each Side

Fig. 26-4 Three-pass vertical-up fillet weld using E-7018 electrode

3. The welding current may have to be set higher because of the greater amount of plate being welded.

4. The beads should be positioned so that the weld is centered in the joint.

Note: All beads should be flat and uniform in appearance and have no undercut or pinholes. This is accomplished by using a proper length arc and good timing when motion is used.

VERTICAL-DOWN WELDING

1. E-6010, E-6011, E-6013 and E-7014 electrodes are all satisfactory for welding vertical down.

2. Stringer beads with no side motion are generally used at higher amperages. See fig. 26-5, 1/8-inch E-6013 vertical-down T joint.

3. Weaves are only used as cover passes with a slight inverted U-type motion.

4. The penetration is not as great when welding vertical down.

No motion required. Hold a close arc.

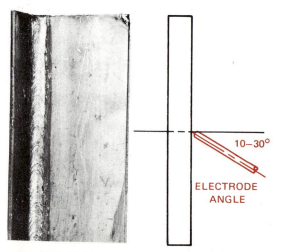

10–30°

ELECTRODE ANGLE

Fig. 26-5 Vertical-down fillet weld using E-6013 electrode and DC straight polarity

REVIEW QUESTIONS

1. When is it an advantage to weld vertical down instead of vertical up?

2. What is the difference in the way the E-6010 and E-7018 electrodes are handled to apply vertical-up stringer beads?

3. When welding vertical up with an E-6010 electrode, does undercut become a problem at the edges of the weld?

4. What electrode is similar to E-6010 in operation and can be used on an AC welding machine?

5. Is the degree of penetration greater with vertical-down welding or vertical-up welding?